British Economic Growth during the Industrial Revolution

Acknowledgements

DURING the writing of this book I was fortunate to be given encouragement and advice by many colleagues and seminar participants too numerous to mention individually; I am grateful to all of them. Ian Gazeley, Mark Thomas, and Tony Venables read parts of the manuscript and made helpful suggestions. I owe a special debt to Patrick O'Brien for his constructive criticism. Most of the book was written during a teaching assignment at Stanford University. I am indebted to University College for allowing me leave and to Paul David and Gavin Wright for the stimulating environment they provided. Barbara Crafts typed the manuscript with great care and made sure that I actually finished it.

Contents

1

Introduction

THE aim of this book is to provide a presentation of recent research on the overall economic growth of the British economy during industrialization which will be accessible generally to undergraduates and of some use to professional economic historians. This is *not* a textbook trying exhaustively to cover all aspects of the industrialization process and the period; the hope is rather to provide a macroframework into which other more detailed microlevel studies can be fitted and which can also act as a background for other investigations.

The main emphasis in the book is to attempt a description of *how* the economy developed, and, in bold outline, this is carried out in Chapters 2 and 3. Questions such as 'why did Britain have the first industrial revolution?' are avoided. Nevertheless, the elaboration of further details of the overall growth process in Chapters 4 to 7 develops a new perspective on some major aspects of the achievements of the economy during early industrialization which will at least offer clues as to why Britain developed as she did.

Most of the famous textbooks available on the British industrial revolution are written without developing much of a comparative perspective of how British economic development relates to the experience of other countries. Chapter 3 puts Britain into the context of development economists' generalizations about patterns of economic development and illustrates that the British experience of industrialization was in many ways atypical not only of recent experience in the Third World but also of European experience in the nineteenth century. A sense of which features of British economic growth and development were unusual may, perhaps, be useful both for those seeking in future to speculate on reasons for Britain's early start in industrialization and for those wishing to analyse the consequences of that early start for latter day economic history. Some speculative ideas about those consequences will be found in Chapter 8.

The major revision to earlier views suggested in Chapter 2 of this book is that growth was considerably slower between 1780 and 1821–30 than was believed by Deane and Cole (1962) and the many writers who have used their findings. Here it is argued that the acceleration of growth was a more gradual process than metaphors such as 'take-off' imply, and that Britain did not achieve growth in real output of 2 per cent per year until after the wars with France, say, in the 1820s—about forty years later than Deane and Cole's figures suggested.

This proposition has important implications for our perception of the process of growth. In particular, productivity growth overall seems to have been only slowly increasing until the second quarter of the nineteenth century and was until then, at the earliest, rapid in only a few sectors of the economy. Moreover, until the 1820s increases in investment struggled to keep pace with greater population-growth pressure. (These points are developed on Chapter 4.) Consequently, as Chapter 5 suggests, consumption grew at best only very slowly before 1830, and this is of great importance in establishing the context for the famous debate on the standards of living of the working classes.

Transfer of productive resources from agriculture into industry and gains in productivity arising from this transfer are major themes in the work of development economists such as Kuznets (1971) and Chenery–Syrquin (1975). The picture emerging from Chapter 3 and further developed in Chapter 6 is that this process was completed very early on in Britain compared with elsewhere. Even by the mid-eighteenth century the proportion of labour force inputs employed in agriculture may have been less than 50 per cent, and by 1840, prior to the repeal of the Corn Laws, there was no gap between average labour productivity in agriculture and in the rest of the economy. Labour was released by British agriculture at an early stage in the development process to a large extent.

This process of structural change was also related, however, to developments in Britain's international trading position. In Chapter 7 it is shown that Britain, relying heavily on exports from only a few sectors, developed a pattern of exports based almost wholly on manufactures when the economy was still at a relatively low income level. Moreover, a substantial part of additional British industrial output was exported during the early years of industrialization, and at the same time the economy started to import significant amounts of food. In other words, Britain increased her specialization along lines of comparative advantage, a process eventually intensified under free trade after 1846.

Before proceeding to this material, it is necessary to make some preliminary points to avoid misunderstanding. First, the unit of analysis is the British economy. Evidently, in some ways this is an arbitrary choice. The British industrialization process was not one in which deliberate government policy decisions played a large part. Recent writers have correctly emphasized that there were important regional differences within the economy, and Pollard (1981, ch. 1) has pointed out that European industrialization in the early nineteenth century was more closely based on natural resource endowments than national boundaries. Nevertheless, there are good reasons why a major level of analysis in economic history should be the national economy. It is this unit for which the most data are available and, in particular, national income figures can be constructed whereas figures for regional incomes do not exist. Further, events such as the financing of the wars with France, perhaps an important reason for the slow growth prior to 1820, were burdens on the national economy, and decisions such as the abolition of the Corn Laws had repercussions on the structure of the whole economy. Finally, the national economy of Britain represented, for many products, a well-integrated national goods market by the early nineteenth century (Granger and Elliot, 1967; Crafts, 1982) and a fairly well-integrated set of factor markets with much more internal mobility of capital and labour than there was in the international economy (Crouzet, 1972; Greenwood and Thomas, 1973).

Second, it should, however, be remembered that the extent of the industrialization of the labour force varied considerably between regions. An initial sense of these differences can be obtained from Table 1.1; this refers to 1841, the year in which relatively complete information on occupations first becomes available in the Census, and a year at the end of the period often referred to as the 'Industrial Revolution'.

Table 1.1 demonstrates quite a wide range of experience; the percentage in manufacturing ranges from 15.4 (Highlands) to 54.3 (Yorkshire, West Riding), and the percentage in agriculture spans from 10.0 (Lancashire) to 62.6 (Highlands). The tendency, of

Table 1.1. Occupational Distributions for 1841 (males)

	% *agriculture*	% *mining*	% *manufacturing*	% *construction*	% *other*	*upperbound % in 'revolutionized industry'*
London	4.1	0.6	41.5	8.8	45.0	18.8
Kent	33.4	0.4	22.5	6.6	37.1	10.3
Surrey	16.3	0.7	33.7	9.1	40.1	13.3
Sussex	45.9	0.3	21.0	7.7	25.2	8.0
Hampshire	37.6	0.2	22.5	7.5	32.2	9.6
Berkshire	44.7	0.2	20.5	6.6	28.1	7.8
Oxon	47.6	0.3	23.9	7.4	20.9	8.0
Bucks	52.5	0.1	20.5	5.6	21.3	5.9
Beds	54.7	0.1	20.5	6.1	18.6	5.4
Herts	46.5	0.2	20.3	7.0	26.0	7.4
Essex	53.1	0.1	18.3	5.8	22.7	8.2
Cambs	53.4	0.1	19.7	5.9	20.8	6.2
Hunts	54.0	0.1	20.4	6.0	19.5	6.3
Norfolk	46.7	0.4	25.2	6.6	21.0	11.8
Suffolk	53.5	0.2	22.3	6.3	17.7	9.0
Cornwall	30.0	27.0	16.4	7.2	19.4	8.1
Devon	38.3	1.6	23.8	8.9	27.5	9.8
Somerset	38.1	3.8	24.2	9.6	24.4	9.4
Gloucs	26.2	3.4	30.1	9.3	31.0	13.6
Wilts	49.2	0.4	21.7	6.9	21.9	11.0
Dorset	44.3	1.3	22.8	9.1	22.5	8.9
Herefords	47.1	0.3	18.3	8.2	26.1	5.7
Shropshire	38.8	10.2	24.9	5.8	20.3	10.2
Warks	21.1	1.4	46.8	7.0	23.7	21.3
Staffs	19.0	13.0	39.5	7.5	21.1	14.3
Worcs	31.7	4.0	37.3	6.0	21.0	14.0
Derbyshire	23.4	9.3	36.5	6.0	24.8	14.0

Notts	28.9	1.9	46.1	4.8	18.2	31.6
Leics	28.5	1.8	44.1	5.3	20.2	29.6
Lincs	55.7	0.2	20.7	6.0	17.4	7.6
Northants	46.9	0.2	29.0	6.4	17.5	7.1
Rutland	56.1	0.1	19.2	6.6	18.0	5.9
Cheshire	22.2	2.6	43.4	6.1	25.7	31.1
Lancashire	10.0	4.1	35.5	7.2	28.6	39.2
Yorks W.R.	14.7	5.6	54.3	6.2	19.2	38.3
Yorks E.R.	42.1	0.4	26.8	6.5	24.2	11.6
Yorks N.R.	47.0	3.0	26.2	7.5	16.3	10.8
Westmoreland	41.6	2.1	29.6	8.3	18.4	14.7
Cumberland	30.0	8.2	31.7	8.3	21.8	18.0
Northumberland	23.1	14.3	30.3	8.5	23.8	15.7
Durham	14.8	20.5	30.8	8.2	25.7	17.4
Glamorgan and Monmouth	18.6	24.0	24.4	7.1	25.8	15.0
North and West Wales	45.5	10.6	19.1	6.7	18.1	6.9
Strathclyde	18.3	6.3	41.8	7.3	26.4	30.7
Dumfries and Galloway	48.9	1.5	25.4	7.8	16.5	13.9
Borders	47.3	0.4	27.2	8.9	16.1	16.8
Lothian	18.9	3.2	36.2	8.6	33.1	15.3
Central and Fife	25.3	6.3	41.5	6.8	20.1	30.7
Tayside	29.9	0.9	42.0	8.0	19.3	31.7
Grampian	51.5	0.5	24.0	6.9	17.1	12.9
Highland	62.6	0.2	15.4	5.8	16.0	6.3
Great Britain	28.5	4.5	33.8	7.3	25.9	19.0

Notes: Derived from Lee (1979). 'Manufacturing' is Lee's categories 3–20. 'Other' includes Lee's 21–27 *and* not classified. Female employment in 1841 was concentrated in domestic service (996,038), textiles (343,600), and clothing and footwear (196,620) out of a 1,803,238 total. The 'upper bound % revolutionized industry' figure for total (that is, male and female) employment is 19.1%.

'Revolutionized industry' subsumes Lee's categories (5) Chemical and Allied Products, (6) Metal Manufactures, (7) Mechanical Engineering, (8) Instrument Engineering, (11) Vehicles, (13) Textiles, and (22) Transport. As the text indicates, this is very much an overestimate of the proportion of employment 'revolutionized'.

course, is for northern and midland areas and south-central Scotland to be the most manufacturing orientated areas, and the south and east of England and the celtic outskirts the least. However, it should also be pointed out that only three areas (Rutland, Lincolnshire, and the Highlands) had a higher proportion of the labour force in agriculture than the 'European norm' for the economy as a whole at a comparable income level (see Chapter 3). Thus, whilst the extent of industrialization varied markedly, the move of the labour force out of agriculture was nevertheless very general by 1841.

Third, and of particular importance, it is customary to refer to an 'Industrial Revolution' taking place in Britain between, say, the mid-eighteenth and mid-nineteenth centuries. The term has often been used without being carefully defined (Hartwell, 1971, pp. 44–5), and has also been applied to smaller changes perhaps only affecting small parts of the economy (Coleman, 1956), and still offers much scope for semantic argument—as the recent interchange between Fores (1981) and Musson (1982) indicates.

If interpreted very literally, then the phrase 'Industrial Revolution' can undoubtedly be extremely misleading. Change in the economy was not, of course, confined to industry during the period in question: changes in the economy had roots in previous periods, and the interpretation of growth presented in this book reinforces the argument that a 'cataclysmic' interpretation of economic change in the late eighteenth century is inappropriate. On the whole, recent research has been tending to stress the gradualness of change when seen from a macroeconomic standpoint, and has also been tending to argue that the 'Industrial Revolution' was not narrowly economic but social, intellectual, and political—at least since Ashton (1948) published. Maybe there would be advantages in the development of a more subtle terminology, but as Ashton noted, 'the phrase "Industrial Revolution" has been used by a long line of historians and has become so firmly embedded in common speech that it would be pedantic to offer a substitute' (1948, p. 2)); economic historians are, in other words, 'locked in' to the phraseology.

That being the case, it is useful to review here two types of economic change which had happened in Britain by the mid-nineteenth century, *both* of which have been labelled as an 'Industrial Revolution'. One kind of economic change might be described as an overall structural change when viewed from an economist's perspective and is the approach and definition adopted by Deane in her famous textbook (1979). The argument looks at Britain in terms of structural characteristics and relates changes in these to the observed generalizations of structural change in economic development made by such writers as Kuznets (1966) or Chenery–Syrquin (1975). As we will see in Chapter 3, by the mid-nineteenth century Britain's labour force had become mostly non-agricultural; there are many signs of an efficient allocation of resources, investment was higher than ever before, the economy was heavily urbanized, and, in effect, Britain was exhibiting many of the features of a 'developed' rather than a pre-industrial economy, and showing them to a greater extent than any other economy had achieved up to that date. In this (Deane's) sense Britain had experienced an 'Industrial Revolution'. Our concern during subsequent chapters will be to establish as much as possible of the magnitude and timing of these changes. The main new result of recent research will be seen as finding that this process of transformation was more gradual than believed by earlier writers and, for instance, had proceeded further than Gregory King perceived by the beginning of the eighteenth century. Nevertheless it is also important to recognize that the structural changes were accomplished at a stage when income had still not risen very much, and had occurred even though overall economic growth had never been very fast.

The second kind of change which could be thought of as representing an Industrial Revolution is associated with technological change. In particular, the picture evoked is of the spread of the factory system and steam-powered mechanization promoting faster growth of industrial output and faster structural change. This sort of definition is one I used in an earlier paper (1977). By this criterion, too, Britain could be said to have had an industrial revolution by 1850, but in this sense the 'Industrial Revolution' was only *part* of the growth and development process.

Indeed, in this view cotton textiles would certainly take pride of place. Even in 1870 use of steam power was to a considerable extent concentrated in textiles, with 580,000 out of 1,980,000 HP (Kanefsky, 1979, p. 373); and as Musson has shown, 'in most industries there was no technical revolution in the century before 1850 . . . traditional handicrafts still predominated in the mid-nineteenth century and that outside textiles (particularly cotton), the primary processes of iron production, and a few other manufactures, there had been no widespread introduction of steam-powered mechanization and the factory system', (1982, pp. 252–3) so that 'the typical British worker in the mid-nineteenth century was not a machine operator in a factory but still a traditional craftsman or labourer or domestic servant' (Musson, 1978, p. 141). Table 1.1 reinforces this point; even with a most exaggerated view of 'revolutionized industry' only 19 per cent of workers would be counted as employed in that category in 1841, and in only a few areas would the proportion reach 30 per cent, whilst in most of southern England it would be well under 10 per cent.

Thus, much of the economic growth experienced by the economy before the mid-nineteenth century could be said to be 'traditional'; as Chapter 4 shows, productivity growth in most sectors was very low. Of course, output in these 'traditional' sectors expanded partly as a response to growth in the 'glamour' sectors, but even before cotton's spectacular development, as Chapters 2 and 3 show, the economy was quite manufacturing orientated and growing. These features would, no doubt, have continued to make for 'revolution' in the sense of structural change even without the superimposition of 'revolution' in the sense of the beginnings of the factory system.

An explanation of the industrial revolution in the second, technological, sense raises very difficult methodological and philosophical questions. As I argued in my earlier paper (1977), economic theory does not provide a suitable covering law to account for such phenomena, theories of invention are essentially stochastic, and primacy in developing the factory system only happened once. In other words it is unlikely that the question 'why was England first?' in terms of the establishment of the factory system in cotton textiles can ever be satisfactorily answered.

On the other hand, economic theory can, at least to some extent, illuminate the experience of the industrial revolution in terms of structural change and the sources of economic growth. Combined with empirical information, the application of relatively simple economic concepts can indicate some of the ways in which the British development process was unique, and perhaps provide at least some clues as to why Britain in the mid-nineteenth century was the richest country in Europe and by far the most industrialized.

2

Estimates of Economic Growth

I INTRODUCTION

THE major pioneering work in which estimates of economic growth over the long run were first produced was that of Deane and Cole (1962). Since its publication their book has deservedly had an enormous influence over the literature of the British industrial revolution. Whilst later publications (Deane, 1968, Feinstein, 1972) have revised and extended Deane and Cole's estimates from 1830 onwards, only recently has their work for the earlier period been the subject of thorough re-examination. This chapter attempts to synthesize the results of recent research with the foundations provided by Deane and Cole and thus to provide new estimates of economic growth in Britain for the period 1700–1830. As such it is an expanded version of parts of a recent paper (Crafts, 1983a).

Right at the outset, however, it is important to emphasize that even though there has been much recent research leading to substantial improvements in our knowledge, estimates of economic growth prior to the mid-nineteenth century for Britain are bound to be in the nature of controlled conjectures rather than definitive evidence.

The major new sources of evidence which have become available since Deane and Cole wrote are as follows. Wrigley and Schofield (1981) have published their series of estimates for the English population for 1541–1871. The picture of population growth which they give differs from that of Brownlee (1926), the series used by Deane and Cole, for the first half of the eighteenth century. There can be no doubt that Wrigley and Schofield's estimates are far superior to those of Brownlee. The change in population growth estimates is of particular interest since Deane and Cole believed that to a considerable extent the growth of sectoral output could be approximated by population growth.

Further important work has been done by Lindert (1980) and

Lindert and Williamson (1982 and 1983a) on the structure of occupations and income distribution. Based on a study of burial records and wage data these authors have been able to revise and improve the pictures of the economy drawn by the contemporary investigators Gregory King for 1688, Joseph Massie for 1759, and Patrick Colquhoun for 1801/3. Probably the most important finding is that Gregory King's picture of occupational structure is biased in the direction of being too agricultural (Lindert, 1980, p. 707), but Lindert and Williamson's work also provides estimates of national income in current prices for the years 1688, 1759, and 1801/3.

There is also much better evidence now available on capital formation, thanks to the painstaking calculations in Feinstein (1978). These will be used to a major extent in reconsidering the sources of economic growth in Chapter 4. Even in this chapter, however, the data provided by Feinstein is substantially useful in enabling improved calculations of income from rents and the growth of construction output to be made.

Finally, a major article by Harley (1982) has made available revised estimates of industrial-output growth. In his article Harley makes a strong case that earlier writers, including Hoffmann (1955) and Deane and Cole, overestimated growth in industrial output between 1770 and 1815. In calculating industrial-output growth prior to 1800 Deane and Cole relied quite heavily on the trade statistics of British exports and imports. Harley's more direct calculations, based on the growth of the use of industrial raw materials and on excise data, suggest that Deane and Cole's use of trade statistics, on occasion at least, led them into error.

In addition, there has been a substantial body of research which contributes to our knowledge of particular sectors' experiences and which can be used to refine the picture of aggregate experience. Thus, we now know more about agricultural productivity thanks to the work on yields per acre of Holderness (1978), Overton (1979), and Turner (1982). Similarly, our knowledge of output in the iron industry has been improved by Hyde (1977) and Riden (1977); and for output in the coal industry Pollard (1980) provides additional information.

Nevertheless, to a considerable extent estimates of economic growth for 1700–1830 depend on the making of plausible assumptions and the use of appropriate statistical procedures. For many sectors we do not have very good information on output, and Deane and Cole were forced to rely on proxy variables for the eighteenth century; for example, assuming that the growth of income from housing was the same rate as population growth. In one very important case, that of agriculture, I have argued on an earlier occasion (1976) that Deane and Cole used implausible assumptions to construct their proxy series.

Elsewhere we run into 'index number problems'. For example, the measurement of industrial output as undertaken by Harley involves the assignment of weights to obtain a weighted average of the growth of individual sectors, and the answer obtained will be sensitive to the choice of weights. For the early nineteenth century the index number problem turns on the measurement of price changes. From 1801 onward Deane and Cole were able to make estimates of income by sector and for the economy as a whole in current values. The period 1801–31 was, however, one of sharp changes in price levels and of the relative prices of different goods: for example, cotton goods were becoming much cheaper relative to other goods. Deane and Cole used measures of price change derived from the Rousseaux price indices to convert their estimates into real income, but it does not seem that this was a very good choice as the weighting of the Rousseaux indices is quite dissimilar to the weights of the sectors in national output (Crafts, 1980b).

The remainder of this chapter is devoted to a detailed consideration of the new evidence, and to attempting to find solutions to some of the statistical problems of estimating economic growth during the 1700–1830 period. The next four sections of the chapter are hard going, and the naïvely trusting or indolent reader will no doubt wish to skip to section VI where the various parts are brought together and the whole is compared with Deane and Cole's results. Self-respecting readers should not allow themselves to avoid the next four sections!

II THE SOCIAL TABLES

The social tables available to us refer only to England and Wales. They are presented here in a revised form resulting from the work of Lindert and Williamson, which has in particular, through an extensive research of burial records, involved correction of the distribution of occupations given by the original investigators.

The social tables have several uses: they represent the best evidence that exists on the structure of the eighteenth century economy for which, of course, there are no censuses; they yield information on the distribution among sectors of the labour force and of incomes; and they can also be used to reveal likely growth rates of the labour force in agriculture and industry, information which will be of particular interest in Chapters 4 and 6. In addition, however, Table 2.1 will also help to make us aware of some of the difficulties of working with eighteenth-century historical sources.

Let us consider first of all the years 1688 and 1759. The categories used by the original writers, King and Massie respectively, do not quite correspond to those used by development economists such as Kuznets who would be particularly keen to divide the economy up into 'agriculture', 'industry', and 'services'. In part this is because of the aims of the writers—Massie was a polemicist wishing to attack the colonial sugar lobby and grouped classes on the basis of tea and coffee drinking; in part the problem is likely to be one of there being less specialization in the eighteenth-century labour force with many individuals working part time in (seasonal) agricultural occupations and part time in traditional industry. Use of the burial record data helped Lindert and Williamson to correct some of the mistakes or odd groupings made by King and Massie, but as Lindert notes, among the many problems of analysing burial records, women were rarely described by occupation and labourers were not usually distinguished by sector (1980, pp. 691–2). Also, the data is given by family rather than by individual, and as a result female domestic service is a category which does not appear, although we know that in the nineteenth century this was a large occupational group (996,038 in 1841 Britain, Lee, 1979). Finally, of considerable importance in considering shares of income by sector, rent does not appear as such; presumably rents, especially agricultural rents, must be represented as a large part of the incomes of the 'High Titles and Gentlemen', but the amount cannot be pinned down exactly.

We must therefore work in terms of the available evidence with caution, making assumptions but considering the sensitivity of any

results we obtain to alternative assumptions. For 1688 and 1759 it must immediately be acknowledged that we do not have information about the proportions of time spent by 'industrial' families in agriculture or 'agriculture' families in industry, or about the differences between categories in terms of wives' labour-force participation, workers other than the head of household, etc. We can only work on the basis of the families of Table 2.1. If we look at Table 2.1 in this manner, in order to make the figures comparable with those derived by Deane and Cole from nineteenth century censuses (see Table 2.2) we would need to omit categories 1 and 9 from the labour force figures, assign categories 7 and 8 by sector, and make an adjustment for domestic service. An upper bound for the proportion in agriculture can be obtained by assigning 7 and 8 in full to agriculture, allowing 10 per cent additional labour-force inputs into domestic service on the basis of nineteenth-century evidence, and omitting 1 and 9. This would give the proportion in agriculture as (16.4 + 20.5 + 22.5)/(100 + 10 − 3.1) = 55.6 per cent. Similarly, in 1759 the result would work out at 48.0 per cent. In fact, these upper bounds may not be all that far from the mark, as we shall discover in a moment.

This calculation immediately reveals a major result of Lindert's work. The upper bound for the proportion in agriculture is 55.6 per cent for 1688. As Lindert himself noted, later writers have been led astray by King, who 'painted a nation consisting of just London and a vast, poor hinterland . . . England and Wales were almost surely more industrial and commercial in King's day than he has led us to believe' (1980, p. 707). By comparison with this recent research Deane and Cole, admitting great uncertainty, talked of a range of 60–80 per cent for the proportion of the labour force in agriculture (1962, p. 137).

1801/3 can best be considered in conjunction with Table 2.2, which reports Deane and Cole's calculations based on the censuses for the early nineteenth century. It is generally agreed that only in 1841 do we get reasonably good estimates of occupational distribution, and Deane and Cole (1962, p. 142) acknowledged that there was a certain amount of guesswork in their early nineteenth-century calculations. Deane and Cole's figures are for Britain, whereas the social table for 1801/3 is for England and Wales. However, for 1841, where we have a good regional breakdown, Table 1.1 demonstrates that the proportions in agriculture and manufacturing in Scotland were very similar to those in Britain as a whole.

Table 2.2. Deane and Cole's Calculations of the Distribution of the British Labour Force in the Early Census Years (%)

	Agriculture, Forestry, and Fishing	*Manufactures, Mining, and Industry*	*Trade and Transport*	*Domestic and Personal*	*Public, Professional, and all others*
1801	35.9	29.7	11.2	11.5	11.8
1841	22.2	40.5	14.2	14.5	8.5

Source: Deane and Cole (1962, p. 142)

For 1801/3 two assumptions can be used with the social tables First, we could treat 1801/3 in the same way as the earlier years to obtain an upper bound figure—this would give 41.7 per cent of the labour force in agriculture. Second, we could assume that category 8 should be distributed between agriculture and non-agriculture on the basis of the proportions suggested by the crude occupational question of the 1831 census, that is, assign only 60 per cent of category 8 to agriculture (Lindert, 1980, p. 711). This gives a lower bound figure of 37.0 per cent—a figure similar to Deane and Cole's estimate for Britain given in Table 2.2. The similarity of these estimates, derived quite separately, gives some reassurance, and especially noteworthy is that our upper bound estimate is only 112.7 per cent of our lower bound figure (41.7/37.0). Given that categories 7 and 8 must surely have been more agricultural in years earlier than 1801/3 this suggests that the estimates for the share of the labour force in agriculture for 1688 and 1759 of 55.6 per cent and 48.0 per cent respectively are unlikely to be radically wrong.

We can now also derive some sectoral labour force growth rates which will be of particular use in Chapters 4 and 6. Using 'upper bound' figures of 55.6 per cent and 48 per cent for agriculture's share in 1688 and 1759 respectively, and regarding industry's labour force as represented only by category 5 in Table 2.1, labour-force growth rates for 1688–1759 per annum are:

agriculture	− 0.05%
industry	0.51%

A moment with a calculator will assure any sceptic that the error possibilities arising from the assignment of categories 7 and 8 would not change these estimates by very much at all, given that the period concerned is 71 years. Calculations for 1759–1801 using the lower bound (upper bound) figure for agriculture in 1801 give labour-force growth rates of:

agriculture	0.06% (0.35%)
industry	1.36% (0.95%)

Finally, in this section the distribution of income between sectors may be considered. In this the allocation of categories 7 and 8, especially in view of the preceding discussion, is of less concern than dealing with the allocation of income accruing to category 1, the 'High Titles and Gentlemen', of Table 2.1.

Deane and Cole in their assessment of growth during the eighteenth century used the following sectoral shares: agriculture, 43 per cent; industry and commerce 30 per cent; rent and services, 20 per cent; government, 7 per cent (1962, p. 78). Table 2.1 suggests that the figure for industry and commerce is distinctly too low, even if one allows that, say, 5 per cent of incomes somewhere in the categories 4–8 of the table should be assigned to domestic service. The figure for agriculture is almost certainly too high unless virtually all the incomes in category 1 are agricultural. But these incomes presumably also receive a substantial proportion of house rents, which were of the order of 6 per cent of national income in the early nineteenth century where Deane and Cole provide information (1962, p. 166). The share of income originating in government varied, particularly with war and peace—the figures in Table 2.1 for 1759 of sectoral shares are almost certainly more representative of 1755 and do not account for the military build-up of the late 1750s. In all likelihood commerce is slightly too high and industry slightly too low since tradesmen such as bakers who were *both* producers and distributors are found in commerce.

The weights chosen for both 1688 and 1759 as likely to represent England and Wales reasonably well are: agriculture 37 per cent, industry 20 per cent, commerce 16 per cent, rent and services 20 per cent, and government 7 per cent. It will be obvious from Table 2.1 that these weights are somewhat arguable; however, as Section VI will indicate, overall growth estimates are not very sensitive to the sorts of variations which could reasonably be proposed. In any case the shares suggested above are not likely to be very far out.

We can compare these figures to those for Britain given by Deane and Cole on the basis of their evidence on current incomes in the early nineteenth century. Shares representative of the early nineteenth century (1801–31) are: agriculture, 26 per cent; industry, 32 per cent; commerce (trade and transport), 16 per cent; housing, 6 per cent; domestic and personal, 6 per cent; government, professional, and other, 14 per cent. The difference from the early eighteenth-century is both striking and easy to remember, agriculture's loss was industry's gain (virtually).

Finally we should note that even in Gregory King's time modern research indicates that England and Wales was already substantially non-agricultural. Even in terms of labour-force allocation only a little over half the labour force seems to have been used in agriculture. As Chapter 3 will show, this feature of the economy was strikingly different from the general European experience.

III INDUSTRIAL OUTPUT GROWTH

Apart from its importance to the measurement of the overall growth of the economy, there is substantial intrinsic interest in examining in detail the growth of industrial output. As we saw in Chapter 1, the phrase 'Industrial Revolution' requires careful thought because there are various definitions of it which can legitimately be made. In exploring the phrase, it became apparent that distinctions could be drawn between the growth of the 'traditional' and 'revolutionized' sectors; a further point which will be documented shortly is that there were very large differences in the rates of growth of the sectors, and consequently changes in their relative importance. Thus, cotton, the most famous fast growing industry, accounted for 2.6 per cent of value added in industry in 1770, but 22.4 per cent in 1831; conversely the leather industry, a slow-growing 'traditional sector', accounted for 22.3 per cent of value added in 1770, but only 8.7 per cent in 1831 (see table 2.3). At the same time cotton goods, relative to leather, had become much cheaper by 1831.

Economists will recognize in the above the elements of a 'classic index-number problem'. Suppose we wish to measure the growth of real industrial output. It is obviously necessary to value the diverse goods produced, and if we wish to avoid the confusion resulting from inflation it will also be necessary to use constant prices—but which prices? In principle we can make quantity indices using many different sets of weights (prices). Consider the following two from the (infinite) set of possibilities.

$$\text{Index } a = \frac{Qa_2Pa_2 + Qb_2Pb_2 + Qc_2Pc_2 + Qd_2Pd_2}{Qa_1Pa_2 + Qb_1Pb_2 + Qc_1Pc_2 + Qd_1Pd_2}$$

$$\text{Index } b = \frac{Qa_2Pa_1 + Qb_2Pb_1 + Qc_2Pc_1 + Qd_2Pd_1}{Qa_1Pa_1 + Qb_1Pa_1 + Qc_1Pc_1 + Qd_1Pd_1}$$

Where Qa_2 is the quantity of good *a* produced in year 2, Pb_1 is the price of good *b* in year 1, etc. There is no reason at all to suppose that the growth in production as measured by the two indices will be the same. (If you doubt this, try some experimental calculations.) In fact, in historical examples we generally find a systematic tendency for the growth as measured by an index like *a* above, for which prices are taken from the later year, to show less growth than an index like *b*, where prices are taken from an earlier year. The intuitive reason behind this is that sectors which expand very quickly generally are able to do so because their products are becoming cheaper (for example, calculators in the 1970s). Thus, care needs to be exercised in the measurement of industrial-output growth if we wish to discuss the appropriateness of the term 'Industrial Revolution'.

It is a problem of this kind that Harley detected in the index of industrial output constructed by Hoffmann (1955) and subsequently much used. Harley points out that for the period 1770–1815 Hoffmann developed an index based on the growth rates of individual sectors, measured by the growth in consumption of physical inputs or the output of physical measures of output, weighted by their shares in value-added. Hoffmann's known sectors included cotton, but he had some sectors for which he was unable to measure growth. He assumed that the missing sectors grew at the same rate as those for which he had information and thus in effect, *inadvertently*, gave a very high weight to cotton textiles, whose growth was exceptional. For further details see Harley (1982, pp. 276–81).

Deane and Cole's approach to measuring industrial-output

growth (1962, pp. 76–9) was different. They distrusted Hoffmann's index and constructed their own 'proxy' for the rate of growth of industrial output in the eighteenth century. In doing so they treated 'industry and commerce' together—that is, their index of output covered not just manufacturing and construction and mining but also transport and trade (that is, distribution/retailing etc.). The index number they obtained for 'industry and commerce' had a 40 per cent weight given to output in 'home industry', with assigned weights of 14 per cent to beer, 22.8 per cent to leather, 2 per cent to candles, and 1.2 per cent to soap output, and the quantities were based on the physical quantities of production subject to excise duty. For the other categories, 'export industry' and 'commerce', accounting for the remaining 60 per cent of the weights, growth of outputs was assumed to be equal to growth in the volume of overseas trade experienced by the economy as a whole. The volume of overseas trade was measured in 'official prices', that is, prices prevailing at the beginning of the eighteenth century.

This procedure is not very satisfactory for several reasons. First of all, it is in effect an *implicit solution* to the index number problem; use of the trade statistics may give an answer which would also be obtained by adopting a particular set of price weights, but even if this is true we do not know what those weights would be and are therefore not in a position to judge their merit. Second, as Deane and Cole themselves stress, the growth of international trade involving Britain was erratic during the eighteenth century (especially when it was affected by war), and particularly so during the 1770s and 1780s (1962, pp. 48–9, 95); moreover, as Deane and Cole's discussion shows, exports and output often grew at different rates even in the three key export trades—woollens, cotton, and iron (1962, pp. 185, 196, and 225). Third, the 'home industry' coverage is rather narrow. Fourth, although, as Table 2.1 showed, the 'commerce' sector was large its treatment is entirely indirect.

For the early nineteenth century Deane and Cole used different information. They were able to form estimates of income in current prices by sector, using data from tax receipts, censuses, and wage statistics. They distinguished 'trade and transport' (commerce) from 'manufacturing, mining, and construction' (industry). They then used measures of overall price change based on the Rousseaux index to deflate their figures for income (output) in current prices into constant prices[1] (1962, p. 170). Unfortunately, the Rousseaux price index numbers are not well suited to this task and, in effect, the outcome is to give a very arbitrary set of weights to the output of different sectors during 1801–31.

Given these difficulties with the existing estimates of industrial output this chapter develops new estimates; in the construction of these estimates it should become clear that, because of the weighting problems discussed above, a range of possible answers exists to the question 'how fast was the growth of industrial output?' In the nature of things this will be true for any period in any country, but in the period under review the range is rather wider than would usually be the case.

A starting point can be obtained from Table 2.3. This table puts together estimates of value added by industrial sector in three benchmark years. Note that 'commerce' is excluded. Obviously not all the sectors of industry are represented, and in some sectors value added may be a little high as not all intermediate goods have been properly accounted for. Nevertheless, the missing sectors cannot be of any very great importance, and the figure for 1801 tallies closely with Deane and Cole's estimate of income originating in manufacturing, mining, and building of £54.3m. (1962, p. 166). As will be clear from the footnotes, the work of Table 2.3 is in any case largely a reconstruction of Deane and Cole's own work.

The important points in Table 2.3 are immediately obvious. Comment has already been made on the changing relative importance of cotton and leather. Perhaps it should be added here that both cotton and iron, the two sectors usually seen as particularly dynamic, were really quite small relative to industrial output as a whole during the late eighteenth century. Also, the rise in building as a sector is interesting. No doubt it reflects both a rising share of investment in national expenditure and the fact that a considerable part of that extra investment went into housing to

[1] In principle, this is simply an indirect method of constructing a quantity index, and indeed this approach has computational advantages and is the most usual way of trying to measure the growth of real output. To see the equivalence define value of output in time t as $\left(\sum_{i=1}^{n} p_i^t q_i^t\right) / \left(\sum_{i=1}^{n} p_i^0 q_i^0\right)$ and define an index of prices as $\left(\sum_{i=1}^{n} p_i^t q_i^t\right) / \left(\sum_{i=1}^{n} p_i^0 q_i^t\right)$. Divide the value index by the price index and the result is $\left(\sum_{i=1}^{n} p_i^0 q_i^t\right) / \left(\sum_{i=1}^{n} p_i^0 q_i^0\right)$ which is a quantity index of the kind discussed in the text.

cope with the added population pressure. In any event, this represents a 'traditional' industry of growing relative importance during industrialization, since, as Musson has noted, 'there was remarkably little technological change in building during this period . . . Comparative lack of technological innovation in building was associated with generally small scale production' (1978, p. 132).

Table 2.4 presents data on the growth of industrial output by sector based on the available evidence of physical quantities consumed as inputs or taxed as outputs in most cases. This material is essentially that used by both Hoffmann (1955) and Harley (1982) to construct their indices of industrial output, although as we have seen it was relatively little used by Deane and Cole (1962). The evidence is presented by decade, except for 1700–60. For that period the excise data is incomplete and the evidence for wool, the single most important sector, consists of widely spaced observations (Deane, 1957). In section VI, as we shall see, it is convenient to treat 1700–60 as a whole for the purposes of growth measurement, and for the moment, in exploring the nature of the 'Industrial Revolution', 1760–1830 is the focal point of our attention in any case. As with Table 2.3, the figures here apply to Britain, whereas the social tables in Table 2.1 were only for England and Wales.

There are quite a number of points of interest arising from Table 2.4. First, and in many ways most important, we see that there is a large dispersion in sectoral growth rates, especially in 1770–1811. Cotton is seen to be an exceptionally fast growing sector, particularly relative to other sectors in 1770–1801. At this time, indeed, there is the appearance of cotton's growth taking place partly at the expense of other textile sectors, and it is noteworthy that the large woollen industry appears as a slow grower.

Second, we see clear signs of growth in industry output prior to the period usually designated as the 'Industrial Revolution', and the growth of 1700–60 is indicative of the continuity which many have seen in the performance of the eighteenth-century economy. In this regard the median sectoral growth rate is an interesting statistic; it has the value 0.67 per cent for 1700–60, varies between 1.32 and 1.65 per cent in the next four decades, and then climbs steadily to a figure of 3.03 per cent in 1821–31. In other words there is a more gradual 'average' advance to set

Table 2.3. Value Added in British Industry (£m., current)

	1770	%	*1801*	%	*1831*	%
Cotton	0.6	(2.6)	9.2	(17.0)	25.3	(22.4)
Wool	7.0	(30.6)	10.1	(18.7)	15.9	(14.1)
Linen	1.9	(8.3)	2.6	(4.8)	5.0	(4.4)
Silk	1.0	(4.4)	2.0	(3.7)	5.8	(5.1)
Building	2.4	(10.5)	9.3	(17.2)	26.5	(23.5)
Iron	1.5	(6.6)	4.0	(7.4)	7.6	(6.7)
Copper	0.2	(0.9)	0.9	(1.7)	0.8	(0.7)
Beer	1.3	(5.7)	2.5	(4.6)	5.2	(4.6)
Leather	5.1	(22.3)	8.4	(15.5)	9.8	(8.7)
Soap	0.3	(1.3)	0.8	(1.5)	1.2	(1.1)
Candles	0.5	(2.2)	1.0	(1.8)	1.2	(1.1)
Coal	0.9	(4.4)	2.7	(5.0)	7.9	(7.0)
Paper	0.1	(0.4)	0.6	(1.1)	0.8	(0.7)
	22.9		54.1		113.0	

Sources:

Cotton: Deane and Cole (1962, pp. 185, 187, 212).

Wool: Deane (1957, p. 220); Deane and Cole (1962, pp. 195–6, 212).

Linen: Deane and Cole (1962, pp. 203–4, 212).

Silk: Deane and Cole (1962, pp. 210, 212).

Building: Following Harley (1982, p. 273) building is taken as the sum of Feinstein's estimates of investment in dwellings, public building and works, industrial and commercial buildings, railways, roads and bridges, canals and waterways, docks and harbours, plus half of agricultural investments (1978, p. 41).

Iron: Value added is computed using prices of pig iron from Hyde (1977, pp. 66, 111, 113, 139–40) and outputs of pig and bar iron together with Harley's conversion factors for bar iron into pig iron equivalents (1982, p. 274). The figure is an overstatement since it really reflects gross output.

Copper: Taken from Lemon (1838, p. 70) and for 1801, Harris (1964, p. 115).

Beer: Derived from Mathias (1959, pp. xxiii, 546) and excise data in Mitchell and Deane (1962, pp. 251–2).

Leather: Derived using the discussion in Deane and Cole (1962, pp. 76–9), and the figures for hides and skins in Mitchell and Deane (1962, p. 266), and price data from Beveridge (1939, pp. 296–7, 457).

Soap: Derived using the discussion in Deane and Cole (1962, pp. 76–9) and the figures for beer 1801, extrapolated backwards and forwards using the excise series for soap in Mitchell and Deane (1962, p. 265) and price data from Beveridge (1939, pp. 145–7, 313).

Candles: As for soap but using the excise series for candles, Mitchell and Deane (1962, p. 262).

Coal: Derived from Pollard (1980, p. 216, 229), adjusted as suggested by Deane and Cole (1962, p. 218), and using price data from Beveridge (1939, pp. 257–8) for 1770.

Paper: Derived using excise figures in Mitchell and Deane (1962, p. 263) together with data on prices and costs from Coleman (1958, pp. 140–1, 169, 203–4).

Table 2.4. *Growth of Real Output in Industrial Sectors (% per year)*

	Cotton	Wool	Linen	Silk	Building	Iron	Copper	Beer	Leather	Soap	Candles	Coal	Paper
1700–60	1.37	0.97	1.25	0.67	0.74	0.60	2.62	0.21	0.25	0.28	0.49	0.64	1.51
1760–70	4.59	1.30	2.68	3.40	0.34	1.65	5.61	−0.10	−0.10	0.62	0.71	2.19	2.09
1770–80	6.20		3.42	−0.03	4.24	4.47	2.40	1.10	0.82	1.32	1.15	2.48	0.00
1780–90	12.76	0.54	−0.34	1.13	3.22	3.79	4.14	0.82	0.95	1.34	0.43	2.36	5.62
1790–1801	6.73		0.00	−0.67	2.01	6.48	−0.85	1.54	0.63	2.19	2.19	3.21	1.02
1801–11	4.49	1.64	1.07	1.65	2.05	7.45	−0.88	0.79	2.13	2.63	1.34	2.53	3.34
1811–21	5.59		3.40	6.04	3.61	−0.28	3.22	−0.47	−0.94	2.42	1.80	2.76	1.73
1821–31	6.82	2.03	3.03	6.08	3.14	6.47	3.43	0.66	1.15	2.41	2.27	3.68	2.21

Sources:

Cotton: Growth rate of retained cotton imports. Deane and Cole (1962, p. 51) and Mitchell and Deane (1962, pp. 177–9).

Wool: Derived from wool clip in Deane (1957, p. 220), and raw wool consumed in Deane and Cole (1962, p. 196).

Linen: Rate of growth of linen yarn imports to 1801 from Deane and Cole (1962, p. 51) and then based on Deane and Cole's linen output index (1962, p. 204).

Silk: Rate of growth of raw silk imports in Mitchell and Deane (1962, pp. 206–7).

Building: Derived following the method described in Table 2.3 but using Feinstein's constant price series (1978, p. 40) for 1760 onwards. For 1700–60 based on imports of timber, Deane and Cole (1962, p. 51).

Iron: Rate of growth of pig iron output derived from Riden (1977, pp. 443, 448, 455).

Copper: Derived from Lemon (1838, p. 70) for Cornwall adjusted for Anglesey output using Mitchell and Deane (1962, p. 151) and Harris (1964, p. 115). For 1700–60, growth is assumed to be at the rate of 1725–60.

Beer: Growth rate of production of barrels of strong beer subject to excise duty, Mitchell and Deane (1962, pp. 251–2).

Leather: Growth rate of excise figures for hides and skins, Mitchell and Deane (1962, p. 266). For 1700–60 growth is assumed to be at the rate of 1725–60.

Soap: Growth rate of excise figures for soap, Mitchell and Deane (1962, p. 265). For 1700–60 the growth rate is assumed to equal that of 1720–60.

Candles: Growth rate of excise figures for candles, Mitchell and Deane (1962, p. 262). For 1700–60 the growth rate is assumed to equal that of 1720–60.

Paper: Growth rate of excise series for paper, Mitchell and Deane (1962, p. 263). For 1700–60 the growth rate is assumed equal to that of 1720–60.

beside the explosive acceleration in the cotton textiles industry.

Third, the index number problem should by now start to become clearer, especially when Table 2.4 is taken in conjunction with Table 2.3. Sectors grow at very different rates during the 'Industrial Revolution' period and, as measured by value-added, their relative importance changed dramatically. Obviously a computation of the weighted average growth rate for 1770–1801, say, would result in quite different answers if value-added weights for 1770 or 1801 were used. (The results would be 1.6 per cent and 3 per cent per year respectively.)

This point can be further enhanced by consideration of Table 2.5. This shows the price relatives for different industries. It should be emphasized that these price relatives refer to the sector as a whole and thus subsume the effects of changes in the composition of output. Also, the years chosen to calculate the value-added benchmarks were not chosen to reveal particular sectors' price histories to optimum effect. The prices are based solely on comparisons of the physical quantities used in Table 2.4's construction and the value-added figures in Table 2.3. These price relatives must *not* be used in other contexts, such as in the construction of cost of living indices to deflate money wages. Table 2.5 is intended only to illustrate the point that relative prices changed during the period of the industrial revolution, and thus to reinforce the idea, already introduced, that the calculation of quantity index numbers for industrial output is difficult for this period. Note that, as one would expect, the 'revolutionized' sectors, cotton and iron, have much lower relative prices in 1831 than in 1770. It must be noted that the lack of detailed price information of a high quality is a handicap in examining industrial-output growth; in particular, it is interesting to note how little information we have on cotton prices prior to 1820.

Table 2.6 exhibits various calculations of the growth of industrial output. Clearly,for some periods the choice of the method of calculation can be quite critical. A description of the calculations will be made first, and then follows a discussion of some of the pros and cons of different measures. Section (a) of Table 2.6 is largely self-explanatory. It should be noted, however, that this method of calculating overall growth rates generally does not correspond to well known indices, although in the circumstances it seems to be the one for which the data are best suited.

Table 2.5. Sectoral Price Relatives (1770 = 100)

	1770	*1801*	*1831*
Cotton	100	123.6	66.2
Wool	100	125.6	113.4
Linen	100	101.0	99.3
Silk	100	200.0	144.4
Building	100	150.0	178.8
Iron	100	59.2	30.1
Copper	100	256.2	129.2
Beer	100	135.3	255.9
Leather	100	129.3	123.8
Soap	100	161.9	115.5
Candles	100	134.5	94.8
Coal	100	116.7	133.3
Paper	100	300.0	193.8

Notes: Derived from Table 2.3 and the sources to Table 2.4. The price relatives are only illustrative of the problem of changes in relative prices. In particular, note that the figures for value-added in Table 2.3 are rounded, and potentially this has a substantial effect on some prices, especially in any year where a sector is small.

Column (4), however, is an approximation of growth based on a 'Divisia' index, which is a well-known one. It is, of course, a form of averaging of the earlier columns. Section (b) of Table 2.6 is based on calculations of the growth shown by quantity index numbers like those illustrated on p. 25. These calculations are more sensitive than the 'Divisia' of column (4) to errors in particular value-added figures. They do show the generally expected price effect; namely, note that growth based on quantity indices using 1770 price weights is measured as much faster than using 1831 price weights, where, in particular, the fast-growing cotton quantities are evaluated at much lower relative prices. The 'Fisher Ideal' represents a compromise, averaged calculation using a geometric average of the 1770 and 1831 price weights results. Note that this calculation in column (8) does produce results very similar to those of column (4), the 'Divisia'.

The choice of which index number to use depends on what question it is intended to answer and what direction of bias is sought: for example, is it desired to obtain an upper bound estimate? The approach to measurement of real income growth which economic historians usually use (although often implicitly) is to work in terms of the utility of a representative consumer. Real

Table 2.6. Various Calculations of Industrial Output Growth (% per year)

(a) *Weighted Averages of Growth Rates of Individual Sectors*

	(1) *Weights Based on 1770 Value Added*	(2) *Weights Based on 1801 Value Added*	(3) *Weights Based on 1831 Value Added*	(4) *Weights Based on Geometric Average of Value Added Adjacent Years*	
1700–60	0.71				
1760–70	1.23				
1770–80	1.79				
1780–90	1.60	3.68		2.40	1780–1801 : 2.11
1790–1801	1.38	2.49		1.83	
1801–11		2.70	2.76	2.72	
1811–21		2.42	2.89	2.63	1801–1831 : 3.00
1821–31		3.54	3.83	3.65	

(b) *Growth Rates from Quantity Indices*

	(5) 1770 Prices as Weights	(6) 1801 Prices as Weights	(7) 1831 Prices as Weights	(8) Fisher Ideal of (5) and (7)
1760–80	1.46	1.40	1.43	1.44
1780–1801	2.34	2.14	1.69	2.00
1801–31	3.37	3.57	2.81	3.08

Sources: Derived from Tables 2.3, 2.4, and 2.6. See text for further discussion.

income can then be thought of as the minimum cost at the prices of the base year of attaining a given level of utility. Figure 2.1 illustrates the argument for those familiar with elementary microeconomics. In figure 2.1 in year 1 a consumer is observed at *P* on indifference curve U_1, and in year 2 this person is at point *Q* on indifference curve U_2. In between, of course, relative prices must have changed, with good *x* becoming relatively cheaper in year 2. In year 1 real income measured at prevailing prices (valuations) in terms of good *x* is *OA*. The evaluation of the quantities consumed in year 2 is *OC* (where *CF* is parallel to *AD*). But the indifference curve U_2 can be obtained more cheaply if relative prices (valuations) are still as in year 1 at *R*. This would give the true measure of real income in year 2 as *OB*, and it is clear that *OC* is an *overestimate*.

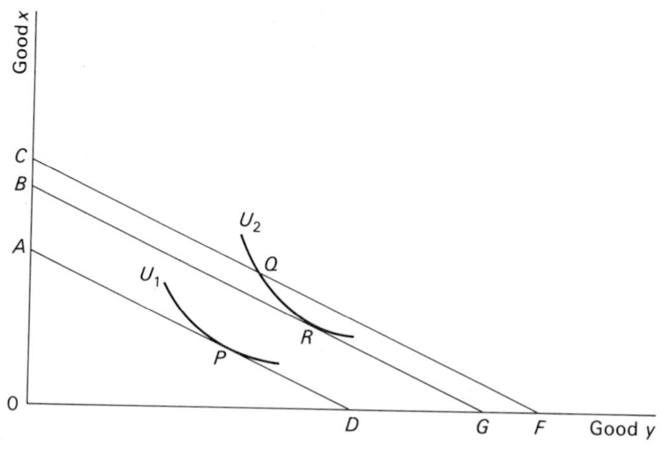

Figure 2.1

The intuitive argument to remember for those who do not like indifference curves is as follows. A quantity index based on weights from a given set of prices generally overestimates the value of consumption in other years relative to the base year, and the reason is that if relative prices have changed then a consumer's consumption will switch towards goods which become cheaper, but in turn these will be evaluated at the margin as relatively less valuable than before by the consumer, whereas the quantity index retains the old relative valuations.

Consider again Table 2.6. The above argument shows that the measures of growth using the perspective of 1770 valuations (that is, having 1770 as the base-year prices for the quantity index) are *overestimates* of the real increase in industrial output measured in welfare terms. Thus, the representative consumer of 1770 would actually value the growth of real industrial output at less than 2.34 per cent per year over 1770–1801, and so Table 2.6's calculation is an *overestimate*.

There is, of course, no particular reason to regard 1770 as a good base year from which to make our evaluations; the choice of base year depends on what question is to be answered. The usual 'ideal choice' made by economists is to evaluate past quantitites consumed in terms of today's prices (relative valuations). As noted above, this preference tends to lead to lower estimates of growth than would be obtained using base weights. Usually the estimates would be made by the indirect method of deflating expenditures by a price index. Nevertheless, as the above discussion has tried to suggest, there is no 'right answer' to the index number problem.

In practice data availabilities dictate a more pragmatic approach and economic historians have shown a preference for avoiding quantity indices using 'extreme' sets of relative prices; that is, growth estimates at the ends of the range of possible calculations. Thus, calculations have often been made on the basis of linking together indices of output using particular price weights or price indices only for fairly short periods and, where the price weights change quickly, using averages. This is exactly the way in which Feinstein proceeded to offer his well-known calculations for real output growth in Britain between 1855 and 1965 (1972, pp. 4, 206–210). I shall adopt a similar procedure and will henceforth use the (Divisia) estimates given in column 4 of Table 2.6, that is, an estimate of industrial output growth of 2.11 per cent per year for 1780–1801 and 3.00 per cent for 1801–31. Obviously, other writers may wish to adopt a different solution to the problem.

Table 2.6 only dealt with 'industry' and not with 'commerce', yet as Table 2.1 indicated commerce was a large part of national income. Table 2.1 also suggests a way to proceed for the eighteenth century. It will be assumed that growth in the commerce sector was at the same rate as national output as a whole. Table 2.1 indicates that income originating in commerce was at a more or less constant proportion of national income in each of the benchmark years. Neither the price evidence from the late eighteenth century (see below Section V) nor that from the United States during industrialization (Gallmann and Weiss, 1969, p. 295) would vitiate the broad conclusion that trade and transport grew with national income in the eighteenth century.

For the early nineteenth century the same assumption, that the growth of commerce is at the same rate as national income, may also not be far off the mark, and Deane and Cole (1962, p. 166) show a fairly constant share of national income in current prices in the trade and transport sector. Their estimate for employment growth in this sector at 1.97 per cent per annum is also very close to that for national income growth in sectors other than transport (see Section VI).

It is possible to obtain some direct evidence on growth of output in the transport sector. For example, from Hawke (1970, pp. 81–2) we learn that real output on the Leeds–Liverpool canal grew at 2.34 per cent per annum between 1801 and 1831, and from excise statistics we find that stage carriage travel grew at 3.88 per cent per annum in the same period (BPP 1857). For both branches of transport the figures cited are probably at the upper end of the overall experience. Baxter's figures (1866, p. 562) give a weight of 56 per cent to water and 44 per cent to road transport in 1834, and using these weights on the above figures for growth of water and road transport would give an estimated growth for transport output of 3.02 per cent per year during 1801–31, a figure which is probably high if anything.

For trade the best assumption seems to be that output grew with employment. This was certainly a traditional sector and could be expected to be an area where labour productivity was pretty much constant. The use of Deane and Cole's employment figures would suggest that employment (output) growth in trade over this 1801–31 period was no more than 1.83 per cent per annum.[2]

[2] Deane and Cole's estimates for employment growth in trade and transport combined are 1.97 per cent for 1801–31 (1962, p. 143). Labour-productivity growth in industry would be at 0.4 per cent per year if output growth was 3.00 per cent per year. Given transport-output growth of 3.02 per cent, and the assumption that productivity growth was no higher than in industry, the predicted growth of employment in transport would be 2.6 per cent per year. Using the separate figure for employment in transport of 0.3 million in 1841 we can then derive the estimate for employment growth in trade of 1.83 per cent per year quoted in the text. It seems likely that labour-force growth in transport is underestimated, and hence employment growth in trade is overestimated, if anything, by this method.

Transport income was about a quarter of trade and transport as a whole in the early 1830s (Baxter, 1866, p. 562; Deane and Cole, 1962, p. 166) and a weighted average of the growth rates suggested above (3.02 per cent for transport, 1.83 per cent for trade) will therefore yield a figure for growth in the trade and transport sector as a whole of 2.13 per cent per year for 1801–31. This may be a little high, but on the other hand it seems unlikely that growth would be less than the 1.97 per cent figure for employment. Section VI indicates that a 2.13 per cent figure would be only just above the overall growth in national output, and this would not be inconsistent with the preceding discussion of eighteenth-century trends.

Evidently the work required to build even crude estimates of the growth of industry and commerce is considerable. Importantly, we have also seen that is necessary to carefully consider the method of converting raw figures into indices of total output. In developing my estimates I have implicitly argued that the procedures used to generate the estimates of industrial output on which standard textbook discussions are based are seriously flawed and exaggerate growth.

With the aid of Table 2.7 we can further compare the new estimates with those of Deane and Cole on whose work the conventional view has been based. It is important to be clear why I do not accept their estimates. We can most conveniently start with 1801–31, the period for which Deane and Cole obtained their estimates by the indirect method of using a price index to deflate figures for output in current prices. In principle, this procedure should be equivalent to constructing a quantity index directly. Unfortunately Deane and Cole used the Rousseaux price index which is of uncertain weighting, does not contain any trade or transport output, and does not represent a basket of goods relevant to finished manufacturing output (Crafts, 1980b, pp. 181–2).

The work of Table 2.6 shows that Deane and Cole's results are implausible; noting that using 1770 price weights to construct an industrial output quantity index gives only 3.37 per cent growth for 1801–31 compared with Deane and Cole's 4.44 per cent. Yet our discussion of index numbers established that the use of 1770 weights would generate an *overestimate* for industrial output growth in 1801–31. Hoffmann's index (which is not flawed for this period) gives 2.8 per cent, rather similar to the calculation in Table 2.6 using 1831 price weights. Finally, note that Deane and Cole's method implies a labour productivity growth in manufacturing of 1.84 per cent per year, which in view of the traditional nature of most manufacturing in this period is surely extraordinary. The compromise estimate I adopted from Table 2.6 gives industrial output growth in 1801–31 at 3.0 per cent per year (and thus a labour productivity growth of 0.4 per cent).

Similarly, with trade and transport Deane and Cole's result again stems from their use of the Rousseaux price index number for which there is no justification. Here the implausibility of the Deane and Cole estimate is best shown by noting that it implies labour productivity growth in this sector, dominated as it was by distribution, to have been 1.05 per cent per year, far higher than the highest possible figure for manufacturing.

Prior to 1801 the differences between the present estimates and those of Deane and Cole are rather less stark than would appear from Table 2.7. The very sharp contrast between 1760–80 and 1780–1801 which appears in the Deane and Cole estimates is to a considerable extent due to the impact of the American War of Independence on the trade figures. In view of the prominence given to the 1780s as a point when industrial-output growth leapt forward, this is especially unfortunate.

Nevertheless, as my earlier discussion argued, the use of international trade data to proxy a major part of industrial growth is in any event undesirable, and consideration of the whole period 1760–1801 still leaves a clear impression that Deane and Cole's estimates of growth are too high. Their implied estimate for labour productivity growth in industry and commerce would be 0.86 per cent per year, much higher than seems reasonable for 1801–31, although no one would doubt that the latter period experienced the faster productivity growth. The present estimates would give labour productivity growth in industry and commerce in 1760–1801 at about 0.2 per cent per year.[3]

[3] This discussion assumes that labour-force growth in industry was 1.36 per cent per year in 1760–1801; Deane and Cole are surely believers in the lower bound figure for the agricultural labour force in 1801, see above section II. For commerce labour-force growth has been taken as the average for 1688–1801 based on the social tables of Table 2.1. For calculations of the present estimate of labour-productivity growth industry and commerce output were taken as equally important, which is about right in 1759 but is no longer true by 1801.

Table 2.7. Comparisons of Estimates of Growth in Industry and Commerce (% per annum)

	(% per annum) *Present estimates*			*Deane and Cole*		
	Industry	*Commerce*	*Industry and Commerce*	*Industry*	*Commerce*	*Industry and Commerce*
1700–60	0.71	0.69	0.70			0.98
1760–80	1.51	0.70	1.05			0.49
1780–01	2.11	1.32	1.81			3.43
1801–31	3.00	2.13	2.71	4.44	3.02	3.97

Sources:
Present estimates: see text.
Deane and Cole: derived from (1962), pp. 78, 166.

It may well be that further research will show that the present estimates are in need of correction. Indeed it is quite possible that my results exaggerate the rate of industrial growth. In particular, it should be noted that Harley (1982, p. 278), using 1841 employment weights obtained a growth rate of industrial output of 1.5–1.7 per cent per year for 1770–1815, whereas my compromise Divisia estimate for the same period is 2.15 per cent per year.

It is clear, however, that Deane and Cole's estimates for the growth of output in industry and commerce during the industrial revolution are too high. Moreover, the acceleration in overall industrial output growth after 1780 was quite modest, as Table 2.6 showed. The implications of these findings for the historiography of the industrial revolution are considerable.

Whilst definitions of 'Industrial Revolution' have tended often to be rather broad and most writers recently have opted for gradualist conceptions of change in the economy as a whole, whether measured in terms of national income, structural change, or seen in qualitative terms, nevertheless the 1780s have still tended to be seen as a point at which there was a large increase (a 'turning point') in industrial-output growth. Thus Ashton's comment that 'after 1782 almost every statistical series of production shows a sharp upward turn' (1955, p. 125) is echoed by writers as diverse as Hartwell (1967, p. 12) and Hobsbawm (1962, p. 28). The more recent textbook discussion of Musson (1978, pp. 64–5) is more equivocal on the upturn of the 1780s, but apparently does not dispute Ashton's claim. Certainly, Deane and Cole's figures for growth in industry and commerce support such a claim, for as Table 2.7 indicates they show growth at 0.49 per cent for 1760–80 compared with 3.43 per cent for 1780–1801.

This picture is nevertheless misleading. It confuses fluctuations from war with longer-term trends, and it over-emphasizes in Ashton's case the experience of a few sectors. Table 2.4 shows that there was indeed a dramatic rise in the growth of cotton textiles, and big upturns in iron and paper, but by no means is the *general* appearance one of a spectacular change in growth. This is borne out by index number calculations from Table 2.6. The weighted average growth rate for 1760–80 based on 1770 weights is 1.46 per cent whilst the Divisia figure for 1780–1801 is 2.11 per cent. If 1770 prices are used to weight a quantity index, growth of industrial output in 1760–80 is 1.46 per cent as compared with 2.34 per cent in 1780–1801, whilst with 1831 prices the former period is 1.43 per cent and the latter 1.69 per cent. (Note that the index number problem prior to 1780 is a trivial one.) Whilst we can talk of a gradual acceleration in industrial-output growth in the 1780s, although only very slight on the 1831 prices quantity index, we surely cannot justify rhetoric such as 'all the relevant statistical indices took that sudden sharp, almost vertical turn upwards which marks the take-off' (Hobsbawm, 1962, p. 28). It is the experience of the revolutionized cotton industry that stands apart.

IV SERVICES OUTPUT

Attempts to measure services output always involve even more serious index number difficulties than we have already encountered, but there is usually much less that can be done in the way of computation. In modern economies much of the output of this sector is not marketed, and the use of the indirect method to deflate figures for current incomes is a hazardous procedure since price indices appropriate to the service sector are hard to devise with the deficient data. Employment growth is thus often used as a proxy measure over short periods, but over longer periods some hypothesis about labour-productivity growth may also need to be made.

Similar problems occur for the industrialization of Britain. Deane and Cole divided services for the eighteenth century into 'government and defence' and 'rent and services', but for the nineteenth century their approach yields estimates for 'housing', 'domestic and personal', and 'government, professional and miscellaneous'. Their estimates are reported in Table 2.8. Our discussion will deal with the non-government activities first.

It must be accepted that estimates of the services sector are the most problematic part of the measurement of real output growth. For this reason some treatments of past economic growth have confined their scope to commodity output—this is Marczewski's (1965) major concern in looking at nineteenth century France and was O'Brien and Keyder's (1978) approach in their comparison of France and Britain's economic growth from 1780–1914. This book does not follow that route simply because services were an important part of British national income—much more so than in France (see Chapter 3). 'Government and defence' and 'rent and services' account for 27 per cent of national output in the sectoral shares established in Section II for the eighteenth century, and 26 per cent in the 1801–31 period.

Table 2.8. Deane and Cole's Estimates of Services Output Growth

(a) 1700–1801	(% per year) *Government and Defence*	*Rent and Services*
1700–60	1.91	0.21
1760–80	1.29	0.68
1780–1801	2.11	0.98

(b) 1801–1831

Housing	*Domestic and Personal*	*Government, Professional, and Miscellaneous*
3.75	3.12	1.97

Source: Derived from Deane and Cole (1962, p. 78 and p. 166).

For the 'rent and services' sector during 1700–1801 Deane and Cole used the assumption that output growth was at the same rate as population growth (1962, p. 77). Because their population figures were for England and Wales their estimates for rent and services growth apply to that area rather than Britain as a whole. For 1801–31 Deane and Cole made estimates of income in current values and then deflated these using the Rousseaux price index (1962, p. 166).

The assumption that rent and services growth matched population growth in the eighteenth century has been retained for my present estimates. No superior alternative is available, in particular because labour force data for the services sector is deficient. As far as rent is concerned, the assumption appears to be reasonable on the basis of Feinstein's work on investment (1978). The labour force data, such as they are, suggest that the assumption may not be too bad for the century as a whole, although it could be off the mark in particular periods.[4] Output estimates must, however, be

[4] We do not have suitable data for measuring labour inputs in services in particular because the social tables do not permit the measurement of domestic service, and burial records do not adequately describe females' occupations. A crude calculation which appears likely to yield an overestimate is as follows. Assume overall labour-force growth is measured by Wrigley and Schofield's

corrected to allow for the new population estimates of Wrigley and Schofield (1981), which were not available to Deane and Cole.

For 1801–31 Deane and Cole's estimates are not acceptable, and indeed they appear to be substantial overestimates. For example, the procedure gives an implied estimate of growth in real output per worker of 1.75 per cent per year in the 'domestic and personal' sector, as high virtually as their own estimates for manufacturing labour-productivity growth given in Section III.

Resort to proxy variables seems likely to be a better way to estimate output for private sector services in 1801–31. For 'housing' I have used Feinstein's estimates of the stock of houses, and have assumed that real output in this sector grew at the same rate as the value in constant prices of the stock of houses (1978, p. 42). Growth at 1.53 per cent is slightly higher than the population growth (1.45 per cent) but is similar enough to offer support to the earlier treatment of eighteenth-century housing. For 'domestic and personal' the growth of output has been assumed to be at the rate of growth of the labour force in that sector—1.37 per cent per year (Deane and Cole, 1962, p. 166). This is a short enough period for us to believe that productivity in this sector would not be rising.

For 1801–31 we also need new estimates for the 'public, professional and miscellaneous' sector. Again it seems that employment estimates are the most promising avenue of approach. Deane and Cole's figures show zero employment growth in this sector of 1801–31 (1962, p. 143) as a result of the decline in government activity after the wars. However, the period was one of considerable expansion in professional employment. Thus,

estimates (1981, p. 529) of the population aged 15–59. Subtract from that the weighted-average growth of those in agriculture, industry, and commerce from the social tables. The result would leave labour growth in the residual services (including government) sector at 1.04 per cent per year for the eighteenth century as a whole, compared with the estimate of 0.56 per cent in output growth. It could then be that output growth in the services sector is undermeasured by a little less than 0.5 per cent per year for the eighteenth century. If this were so, the impact on measured national-income growth would be to raise it by less than 0.1 per cent per year. This calculation surely overestimates the growth in private-sector services labour because the government military sector was much bigger at the end of the century than the beginning. Allowing for the increase in military personnel would suggest that private-sector services labour grew at only 0.74 per cent per year over the eighteenth century, only a little higher than 0.56 per cent output growth estimate. Thus it is probable that following Deane and Cole's assumption of estimating private-sector services growth by population growth will not lead to any serious errors for the century as a whole.

although employment overall did not grow, there was an expansion of skilled and a contraction of unskilled labour. According to Williamson (1982, p. 48), a lawyer on average received about 4 times the pay of an unskilled government worker. If we take this to represent relative productivity, and use data also from Williamson on the growth of different occupational categories, then we get an estimate for the growth of 'real employment' of 1.37 per cent per year.

For the eighteenth century Deane and Cole estimate expenditure on government and defence using figures of net government expenditure on goods and services deflated by the Schumpeter–Gilboy price index. These have been retained for the present estimates. Again there is no reason to believe that this price index is appropriate. Unfortunately, employment data are not good enough to replicate our nineteenth-century approach for the eighteenth century. On the other hand, price changes for most of the eighteenth century were gradual and the problem of using the Schumpeter–Gilboy price index may not be severe, except perhaps for the very end of the century. In Section VI estimates of national-income growth for the eighteenth century both with and without 'government and defence' are presented on account of both the difficulties of dealing with this sector and its volatility between war and peace.

Table 2.9 gathers together the preceding material into new estimates of services sector output growth.

Table 2.9. Revised Estimates of Services Sector Output Growth

(a) 1700–1801	*Government and Defence*	*Rent and Services*
1700–60	1.91	0.38
1760–80	1.29	0.69
1780–1801	2.11	0.97

(b) 1801–1831		
Housing	*Domestic and Personal*	*Government, Professional, and Miscellaneous*
1.53	1.37	1.37

Source: See text.

V AGRICULTURAL OUTPUT

The view of agricultural change that has been generally adopted following the research of the past 25 years has been well summed up by Jones (1981, p. 85):

> In the main the productivity gains of eighteenth-century agriculture were made by extending techniques pioneered, and more than merely pioneered, before 1700, within a set of institutions also brought into being then . . . The concept of an 'agricultural revolution' is not appropriate for such a set of changes.

This view is quite different from the previous prevailing orthodoxy which emphasized the importance of heroic individuals, placed the important changes in output and technique concomitant with (and much enhanced by) parliamentary enclosure, and is associated especially with the writer Ernle. Woodward (1971, p. 323) neatly sums up this older orthodoxy:

> For Ernle the agricultural revolution was essentially parallel to the industrial revolution falling within the period 1760 to 1830 or thereabouts. English farming in the early eighteenth century was portrayed as wallowing in a state of medieval backwardness; indeed 'Farmers of the eighteenth century lived, thought, and farmed like farmers of the thirteenth century.'

The discarding of Ernle's views has been based on the accumulation of many detailed microlevel studies, but by their very nature these tend to talk about the diffusion of technical changes rather than being able to discuss output growth.

Indeed, for the eighteenth century there are no data for aggregate agricultural output, and estimates of agricultural growth have to be made by inference. Deane and Cole estimated agricultural-output growth on the basis of population growth with a very small adjustment for changes on net grain exports (1962, pp. 65, 74). The results they obtained show growth in agricultural output of 0.24 per cent per year between 1700 and 1760, and 0.56 per cent per year between 1760 and 1801. Since the population figures are for England and Wales rather than Britain, the estimate is of course for England and Wales only as was the case for eighteenth-century rent and services.

Deane and Cole's technique is not a good one, however, as Cole (1981) has subsequently accepted. We would expect consumption per person of agricultural goods to be affected by changes in relative prices and by changes in per capita incomes. We know that the relative price of agricultural goods was falling during 1710–45 and rising sharply from 1760 to the end of the century (Deane and Cole, 1962, p. 91). Everyone agrees too that per capita incomes were rising during the eighteenth century, and there is evidence to suggest that the income elasticity of demand for food was quite high (Crafts, 1980a), as indeed experience in contemporary Third World countries also suggests.

Using our knowledge of the determinants of the demand for food can help to solve the problem of measuring agricultural-output growth. In particular, we can deal relatively easily with a period like 1700–60. At the beginning and end of this period relative prices for agriculture were the same (O'Brien, 1982a, Table 1; Deane and Cole, 1962, p. 91) and we also know that net imports were of negligible importance. The rate of growth of home output would have been equal, then, to the rate of growth of demand for agricultural goods. The rate of growth of home demand would be equal to the rate of growth of the population plus the growth coming from higher per capita income. This last would depend on the percentage rate of increase of income per head multiplied by the income elasticity of demand for agricultural food. (Income elasticity is defined as the percentage increase in quantity demanded divided by the percentage increase in income level, holding prices constant.) On the basis of my earlier paper (1980a) the income elasticity of demand for agricultural goods is taken to have been 0.7. Thus, symbolically:

$$\Delta Qag \,/\, Qag = \Delta pop \,/\, pop + 0.7\,(\Delta Y \,/\, pop)/(Y \,/\, pop)$$

Again in calculating the result of this formula Wrigley and Schofield's (1981) population growth-rate figures have been used.

In evaluating the rate of growth of income per head we have to take account of our information on the other sectors of the economy already discussed (industry, commerce, and services) and their importance in national income. Obviously, agricultural income itself must be allowed for in national income growth, and the estimate obtained finally for agricultural output growth is the outcome of a tedious iterative calculation. For 1700–60 the estimate turns out to be 0.60 per cent per year. (A summary of this and the estimates for later periods is given in Table 2.10.) It is reasonable to assume that within this period the fastest agricultural growth would be before 1740 since agricultural prices were falling then, but exact periodization is perhaps best avoided since errors in measured growth rates will be larger over short periods and the data for many components of industrial output in 1700–60 are not well suited to the computation of short-period growth.

Let us now consider 1760–1801. We can proceed for this period in two ways—using essentially the same method as for 1700–60 but allowing for imports, which started to matter after 1780 especially, and for price changes, or by employing the indirect approach to quantity measurement discussed in Section III—using price indices to deflate money incomes for agriculture. Both methods will be used, starting with the second one.

Based on Section II's discussion of sectoral shares and agricultural growth for the period 1700–60, together with Lindert and Williamson's estimate for the national income of England and Wales in Massie's year, a figure of £25.2m. for agricultural output in England and Wales in 1760 can be estimated. Deane and Cole's figure for agricultural income in Britain in 1801 is £75.5m. This would be a figure of 200/232 × 75.5 = £65.1m., if multiplied by the ratio of the national incomes (Deane and Cole, 1962, p. 166; Cole, 1981, p. 65). O'Brien offers us a good price index for agricultural goods for this period, and using his smoothed series, which is appropriate since benchmark years are supposed to 'represent their period', we can obtain a 'normal' figure for England and Wales's agricultural income in 1801 at 1760 prices of £30.1m., (O'Brien, 1982a, Appendix 1). This indicates a rate of growth of home agricultural output of 0.44 per cent per year—a rate of growth considerably below population growth and lower than Deane and Cole's 0.56 per cent estimate. An allowance for agricultural imports based on the work of Davis (1979, pp. 102–14 and 1969, pp. 119–20) would make agricultural consumption £28.2m. in 1760 and £34.4m. in 1801 (both in 1760 prices), and consumption would therefore have grown at 0.50 per cent per year.

The application of the demand approach gives a result consistent with this figure, although the application of the approach is very tedious in view of the price changes of the period, and somewhat hazardous because price information for sectors other than agriculture is not good. The outcome of the calculations would give growth of demand at 0.97 per cent per year without allowing for price changes affecting demand; but with agricultural prices rising relative to other prices at an estimated 0.59 per cent per year demand growth would be reduced to 0.50 per cent per year.[5] If the demand approach is applied to the two sub-periods used for industry, which is hazardous as we do not have money income figures for agriculture in 1780 as a check and because shorter periods raise the risks of faulty estimates, the results are that agricultural output growth would be at 0.13 per cent for 1760–80 and 0.75 per cent for 1780–1801.

For 1801–31 Deane and Cole gave estimates for agriculture for Britain based on figures for current incomes deflated by the Rousseaux agricultural price index. This index is much less problematic than their use of Rousseaux price indices for other sectors. Nevertheless the present estimates were based on O'Brien's carefully weighted price index extended to 1831.[6] The resulting growth rate is 1.18 per cent per year, which is comparable to Deane and Cole's figure of 1.2 per cent for 1801/11–1831/41 (1962, p. 170). Applying Deane and Cole's procedure to 1801–31 yields 1.64 per cent, however. Given that the present approach takes 1801 as a 'representative year' we should probably regard the present estimates as essentially consistent with Deane and Cole's.

[5] When prices are changing we have two problems in measuring agricultural-output growth by the demand approach: estimating the change in relative prices, and estimating the impact of the change in prices on demand. In estimating price changes we need to examine price changes for goods other than agriculture, and data are poor—especially for services. The best procedure available is to estimate changes in other sector's prices and to compare growth in current incomes by sector between 1759 and 1801 using Table 2.1 with the new estimates of growth at constant prices. The results give industrial prices rising at 0.52 per cent per year, commerce and services at 1.86 per cent per year, which together with the evidence for agriculture leads to an estimate that relative prices of agricultural goods rose at 0.59 per cent per year. (The procedure also yielded an overall inflation rate of 1.52 per cent per year and commerce inflation of 1.47 per cent. This rough constancy of the relative price of commerce appears consistent with my treatment of this sector in Section II.) The sensitivity of demand to price changes depends on the price elasticity of demand; this would seem likely to have been slightly higher and of opposite sign to the income elasticity of demand for agricultural goods, and thus a value of -0.8 may be plausible (Crafts 1976, p. 230). The effect of price rises of 0.59 per cent per year would thus be to reduce demand growth by $(0.8 \times 0.59) = 0.47$ per cent per year.

[6] The extension of O'Brien's index to 1831 was based on data taken from Mitchell and Deane (1962, p. 499, 495) for wool and grain prices, and from Beveridge (1939, p. 425, 426, 429) for milk, pork, mutton, and beef. Weights of 60 per cent and 40 per cent were given to arable and meat respectively.

Nevertheless, the 1.18 per cent figure is probably one of the more doubtful estimates in this chapter. The 1831 current income for agriculture is probably somewhat questionable, as 1831 is the year furthest away from income tax data prior to the reintroduction of income tax in the 1840s. Moreover, the demand approach would predict a higher growth of perhaps 1.88 per cent per year. The discrepancy may partly be that, because of changes in income distribution, the demand approach overestimates.[7] The potential impact on national income growth would be to raise the figures given in Section VI by 0.18 per cent per year, if the 1.88 per cent figure were adopted.

The new estimates for agricultural output growth are collected together in Table 2.10, where Deane and Cole's estimates are also exhibited for comparison.

Table 2.10. Estimates of Agricultural Output Growth (%)

	Present Estimates	*Deane and Cole*
1700–60	0.60	0.24
1760–80	0.13	0.47
1780–1801	0.75	0.65
1801–31	1.18	1.64

Source: See text.

It is now useful to explore slightly more the connection between these growth estimates and research on agricultural change. In particular, this may be desirable because at first sight it may seem strange that the initial phase of the parliamentary enclosure movement appears to coincide with lower growth in 1760–1801

[7] Lindert and Williamson (1983a, p. 98) show a significant shift in favour of the top 10 per cent of income receivers, who probably had a much lower income elasticity of demand for food, occurring between 1801/3 and 1867. Note that substituting 1.88 per cent growth in agriculture would not actually resolve matters for the following reason. Using the same methodology as in footnote 5, agricultural prices were taken to be falling at 0.5 per cent per year, but this was based on a 1.18 per cent growth in agricultural real output. To be consistent, further changes in prices would now need to be put in the demand equation if 1.88 per cent is the assumed rate of output growth; *but* this rate of growth *itself* would then be too low to be consistent with the now revised estimate for demand growth. The iterations do not converge quickly. It seems probable that the difficulties come both because the income elasticity figure of 0.7 is too high for 1801–31 and because Deane and Cole's figure for agricultural income in 1831 is in error.

than before. However, whilst this might have seemed strange to Ernle, it is perhaps quite plausible now.

The main gains in agricultural productivity came from the introduction of cropping systems which Jones suggests was occurring more quickly before 1760 than in 1760–1800 (1981, p. 85). Allen has provided evidence suggesting that one of the main results of the early phase of parliamentary enclosure may have been to redistribute revenue to landlords, rather than to increase output (1982). Turner (1982, p. 506) believes that the main growth of agricultural land productivity came before 1770, and we know that the land used in agriculture rose only slightly during the eighteenth century—Jones (1981, p. 70) puts the figure at 5 per cent for the whole century. We do not have investment figures prior to 1760, but Feinstein (1978, p. 41) puts real investment in agriculture in the 1790s at about twice the levels of the 1760s, evidence which might tend to match the relative growth suggested for 1780–1801 relative to 1760–80.

It would seem that to a large extent agricultural growth came from increases in productivity rather than factor inputs; although eventually there was a significant volume of investment, the level of the 1790s was essentially sustained to the 1830s—land inputs rose only slowly even during the French wars and labour inputs were never fast growing. The best estimate given above for agricultural-labour growth was −0.05 per cent per year for 1700–60, 0.06 per cent for 1760–1801, to which can be added a figure of 0.2 per cent per year for 1801–31 based on Deane and Cole (1962, p. 143).

It is therefore of interest to consider what evidence there is on cereal yields. Arable was about half the output of English agriculture during the eighteenth century, and whilst, of course, discrepancies between the behaviour of cereal yields and our estimates of agricultural productivity could be accounted for in the livestock side of agriculture, on the whole it would seem to be likely that arable yields should be consistent with the new estimates of Table 2.10.

Cereal yields have been controversial, as Turner's summary admirably demonstrates (1982). Nevertheless, there seems to be a distinct possibility that the evidence for wheat, the most studied crop, matches Table 2.10 to a reasonable degree. Calculations have been made at the macro and micro levels. The most careful macro calculations are provided by Bennett (1935) and Turner (1982). On the basis of Bennett's data it seems reasonable to conclude that wheat yields were around 11 bushels/acre in 1688 and 17 bushels/acre in 1760. Turner indicates that in the late 1790s a figure of 19.5 would be appropriate. We can add to this that by 1850 yields appear to have risen again to about 27 bushels/acre, based on Lawes and Gilbert (1880, p. 330). These calculations do seem to be essntially consistent with the present estimates.

Bennett's figures, which have long been doubted because of a rival calculation by Fussell (1929), appear to have been substantially vindicated by recent microlevel research by Overton. Overton's work is based on East Anglia where we might expect yields to be a little higher than the average (they were about 20 per cent higher in the 1790s; Turner, 1982, p. 506); he shows on the basis of detailed work with probate records that yields were around 12.5 bushels in the 1670s and rose sharply thereafter by perhaps 33 per cent between the end of the century and 1735 when his data end. Thus East Anglia in the mid 1730s is similar to Bennett's average estimate for 1760. Comparing Overton's figures with Turner's from the Crop Returns for the 1790s for the same area suggests that the rate of advance was only about half as fast between 1735 and 1800 as for 1700–35.

Our overall picture therefore has some credibility in terms of recent research into agricultural change. It suggests that for the eighteenth century the fastest growth of agricultural output occurred before 1760. This growth was surpassed, indeed probably doubled in the first 30 years of the nineteenth century as agriculture then became more capital intensive. Throughout, the increase in labour inputs was probably very small when measured in terms of persons in agriculture.[8]

VI GROWTH IN NATIONAL OUTPUT

We have now discussed each of the sectors of the economy and are in a position to consider national output as a whole. In doing so we should remember that the estimates essentially refer to England and Wales for the eighteenth century and Britain for 1801–31. Even this statement needs elaboration, however, because of the

[8] As the discussion in Chapter 6 indicates, it seems very probable that hours worked per agricultural worker per year rose during 1700–1831.

difficulties with the data we discussed in the previous four sections. Strictly speaking, for the eighteenth century services are based on English population growth, agricultural estimates are for England and Wales, and industrial-output growth is for Britain. It may be that the picture given in Table 2.11 for the eighteenth century is also representative of Scotland, after all, employment structures were similar in the nineteenth century as we have seen, but this remains to be discovered by future research.

Table 2.11. Estimates of Growth of National Product

	Present Estimates		*Deane and Cole's Estimates*	
	National Product	*National Product per Head*	*National Product*	*National Product per Head*
1700–60	0.69	0.31	0.66	0.45
1760–80	0.70	0.01	0.65	−0.04
1780–1801	1.32	0.35	2.06	1.08
1801–31	1.97	0.52	3.06	1.61

Source: See text.

Most of the details necessary to construct the present estimates in Table 2.11 have already been discussed in the preceding sections, but a few more minor points need to be made. In particular, sectoral shares for the periods after 1760 need to be mentioned. For 1801–31 Deane and Cole's estimates (1962, p. 166) give the average sectoral shares of agriculture, 26 per cent; industry, 32 per cent; trade and transport, 16 per cent; housing, 6 per cent; domestic and personal, 6 per cent; government, professional, and other, 14 per cent. The sectoral shares for the period 1780–1801 based on the sectoral growth estimates suggest agriculture at about 25 per cent, industry at 32 per cent, and the other sectors as for 1700–60. National product growth is then taken to be the weighted average rate of growth of the individual sectors. Since their weights only change very slowly the index number problems involved here are small.

Table 2.11 merits some discussion. First of all, however, a warning is required. We have seen that there are considerable data problems in looking at national-output growth; Table 2.11 is essentially a set of hypotheses. It is therefore not sensible to try to pinpoint turning points, and calculations of growth rates over periods as short as a decade are definitely not very reliable. Moreover, the periodization is essentially an outcome of the exigencies of the data, and more than that should not be read into the table. In many ways I would have preferred to leave 1760–1801 as an undivided period, but in view of the historiographic interest in 1780–1801, and because my disagreements with Deane and Cole's estimates are most pronounced for the post 1780 period, I have given estimates for 1760–80 and 1780–1801.

We can check first of all that the pattern of acceleration given in Table 2.11 is not an artefact of the behaviour of government and services, where admittedly I had to resort to proxy variables which others might prefer not to use. Table 2.12 provides evidence on this point. 'Commodity output' is simply a weighted average of the growth of agriculture and industry, and the omission of the government sector is a calculation which simply involves reweighting the other sectors accordingly. It is immediately apparent that the pattern of national-output growth in Table 2.11 is essentially replicated in both columns of Table 2.12.

Table 2.12. New Estimates for Growth of Commodity Output and Private Sector Output (% per year)

	Commodity Output	*Private Sector Output*
1700–60	0.64	0.60
1760–80	0.61	0.66
1780–1801	1.35	1.26
1801–31	2.18	2.07

Source: see text.

The main features of the new estimates compared with those of Deane and Cole can be summarized as follows:

(i) For the period as a whole per capita real-income growth is lower in the new estimates by 0.4 per cent per year. This is important in that it suggests both that there was less scope for consumption to rise and less acceleration in productivity growth than Deane and Cole's estimates indicated. These points are developed in Chapters 4 and 5.

(ii) Agricultural growth is slower after 1760, and faster before 1760, than Deane and Cole estimated. This seems consistent with the tendency of research in agricultural history since Deane and Cole wrote, namely to further play down the achievements of the parliamentary enclosure era and to continue to find evidence of progress in the century or so before 1760.

(iii) The acceleration in growth is much more gradual in the new estimates. In particular, it seems clear that the economy did not reach 3 per cent per year growth in real output before 1830. Moreover, there is further reason after looking at Table 2.11 to feel that dramatic descriptions of the 1780s are inappropriate. It should be noted that the Divisia index of industrial output does not show growth over 3 per cent per year until the 1820s.

(iv) The sustained growth of commodity output over 2 per cent per year was probably not achieved until the 1820s. I have already argued that calculation of growth rates over short periods is particularly vulnerable to error, and it would definitely be unsafe to so examine proxy variables. For what it is worth, commodity-output growth would be estimated at 2.34 per cent for 1801–11, 1.72 per cent for 1811–21 (where agriculture grows very slowly), and 2.50 per cent for 1821–31. The same conclusion would probably be valid for real national output overall; namely that sustained growth at over 2 per cent per year was delayed to the 1820s.

These results have many implications which later chapters will explore. Straight away, however, they can be used to throw further light on the literature which seeks to define the 'Industrial Revolution', since growth of national output is obviously a part of the broad definition. As such the new estimates of national-output growth add further to the arguments for abandoning the view that at the macrolevel we are going to find a marked discontinuity. Before commenting further on the notion of Britain's 'Industrial Revolution', however, we must further consider structural change, and Chapter 3 is devoted to the description of this.

3

Patterns of Development: Britain in a European Context

I INTRODUCTION

In Chapter 1 we came across the idea that the 'Industrial Revolution' can be defined and interpreted in terms of structural change in the economy. In fact, such changes are an important part of the way both Deane (1979, ch. 1) and Mathias (1983, ch. 1) interpret the idea of 'Industrial Revolution'. Similarly, Kuznets examines Britain, and other European countries in the nineteenth century such as France, Germany, Sweden, etc., as examples of 'modern economic growth'. In so doing Kuznets (1971) was searching for evidence of systematic patterns of characteristics of economies as they grew and became richer. Obviously such a search could help to generate inductive statements to do with the causes of economic development. A further development of this way of looking at economies came from Chenery–Syrquin (1975) who mounted a large statistical exercise from which they produced a stylized description of 'Normal Variation in Economic Structure with Level of Development' in the post World War II period.

The task of this chapter is to place Britain's experience between 1700 and 1910 firmly into the context of other countries' experiences of economic development. In particular, special emphasis will be put on the experience of other European countries prior to World War I. From this approach it will be possible to discuss the timing of Britain achieving the characteristics of a developed country, and Deane and Mathias's conceptions of Britain's 'Industrial Revolution' will be given a firmer empirical basis of support.

It also emerges during this chapter that Britain's transition to economic development was not at all typical of nineteenth-century European experience. This theme has been coming to the fore in comparative economic history. Thus O'Brien and Keyder state that

Economic theory lends no support to assumptions, present all too often in writings on French backwardness, that there is one definable and optimal path to higher per capita incomes and still less to the implicit notion that this path can be identified with British industrialization as it proceeded from 1780 to 1914. (1978, p. 18)

In this book, however, the argument that Britain's development differed from that of other countries is explored with a view to gaining insights both about the nature of Britain's early start in industrialization and some of the consequences of that early start. Thus, this chapter also provides a starting point for the more detailed discussions of Chapters 5 to 8.

II TRANSITIONS TO DEVELOPMENT

The work of Chenery–Syrquin (1975) provides a very helpful starting point for an examination of structural change during economic development. Table 3.1 gives an account of the typical characteristics of countries at different levels of income during the period 1950–75. The presentation has been adapted somewhat from the original, in particular to allow for the problems of measuring incomes in poor countries *vis-à-vis* rich countries.

Table 3.1 can be thought of as giving an estimate of the *average* structure of economies at different income levels. The estimates were produced by regression methods and, of course, the equations used did not by any means fit the data perfectly, although in general the degree of explanation was quite good. (For details, see Chenery–Syrquin, 1975, pp. 30, 38–9, 48–9). It should be borne in mind therefore that quite a lot of dispersion about the 'normal' behaviour of Table 3.1 is observed in practice.

The picture of development to be gleaned from Table 3.1 is a very conventional one. It may be of some help to pinpoint the countries in 1970 which were at the income levels used in Table 3.1. Based on Kravis *et al.* (1978, pp. 232–6) examples would be: at $300, India; at $400, Sierra Leone; at $550, Congo; at $700, El Salvador; at $900, Iran; at $2,300, Italy. Alternatively, looking ahead slightly to Table 3.2 we find that $900 represents an income level comparable with that of Britain in 1870 or France in 1910, whilst $300 would be below Britain's level in 1700 and rather like Sweden's in 1860.

For the countries of 1950–70 we seek striking contrasts in the

Table 3.1. The Development Transition, 1950–70

	Per Capita Income, 1970, US Dollars					
	$300	*$400*	*$550*	*$700*	*$900*	*$2,300*
Crude Birth Rate	44.8	42.5	38.8	35.8	32.6	19.1
Crude Death Rate	19.0	17.0	14.1	12.5	10.9	9.7
Urbanization	20.5	26.4	34.0	40.0	46.2	65.8
Percentage of Labour Force in Primary Sector	66.7	62.7	57.3	52.4	46.6	15.9
Percentage of Labour Force in Industry	8.9	11.4	15.3	18.5	21.9	36.8
Percentage of Income in Primary Sector	46.3	41.3	34.6	29.7	24.9	12.7
Percentage of Income in Industry	14.5	17.0	20.5	23.3	26.2	37.9
Consumption as % of National Expenditure	73.0	70.9	69.1	67.7	66.1	62.4

Investment as % of National Expenditure	15.4	16.7	18.3	19.5	20.8	23.4
Government Spending as % of National Expenditure[a]	13.4	13.6	13.5	13.4	13.5	14.1
Foreign Capital Inflow as % of National Expenditure	2.4	2.1	1.7	1.4	1.1	0.1
Exports as % of National Expenditure	19.1	20.2	21.4	22.4	23.4	24.9
Manufactured Exports as % of Total Exports	9.2	11.5	15.5	17.8	21.6	52.6
School Enrollment Ratio[b]	0.354	0.429	0.522	0.592	0.663	0.863

Notes:
[a] Spending on goods and services.
[b] The fraction of those aged 5–19 enrolled at primary or secondary school.
Source: Based on Chenery–Syrquin (1975, Table 3). For full definition of all variables and their sources see Chenery–Syrquin (1975, pp. 180–7). In this presentation the income variable is defined in 1970 US dollars based on 'purchasing power parity' using Kravis *et al.* (1978) to adjust Chenery–Syrquin's income values. Interpolation has been used where necessary. Simulated values are for a country of 10 million people—Chenery–Syrquin's equations include terms in both the logarithm of population size and the logarithm of the square of population size.

characteristics of a $2,300 country and a $300 country. The former would have low birth and death rates and moderate population growth; the latter a high birth rate and rapid population growth. The $2,300 country is highly urbanized, has only a small share of the labour force and national product in agriculture, has half its exports made up of manufactures, and invests nearly a quarter of its national income. By contrast the $300 country has only about a fifth of its population urbanized, has a vast majority of the labour force in primary production which in turn accounts for the vast majority of its exports, has a large productivity gap in the sense that the proportion of income originating in agriculture is much smaller than agriculture's share of the labour force, and invests only about a sixth of its income whilst experiencing a considerable inflow of foreign capital.

It may be worth also considering the position of a $900 country. This income level is clearly well into the transition and, as already noted, represents a prosperous country by the standards of the pre-World War I era. The experience in 1950–70 suggests that such a country would be very different from a $300 country. In particular, note that a $900 country on average would have become much less rural, would be industrialized in terms of receiving slightly more income from the industry sector than from primary production, would already have increased its investment by over 5 per cent of national income, and would have started to develop a significant amount of manufacturing exports. In other words, in the broad sense of the term, an 'Industrial Revolution' would have occurred.

Obviously we must be cautious in our interpretation of Table 3.1. Remember that individual countries will often diverge from this average picture; for example, consider the experience of certain oil-rich economies in the recent past. Also, *within* countries different areas can be expected to have substantially different degrees of urbanization and industrialization. Moreover, countries have different export specializations as, for example, their natural resources differ.

Table 3.1 only conveys part of what economic development involves; it dwells on measurable characteristics and on economic variables. Obviously societies change as income rises. Table 3.1 does not deal with cultural, political, or sociological change, all of which we know to be an important part of the development process. Nevertheless it does provide a helpful framework of reference with which we can now approach the industrialization experience of nineteenth-century Europe.

Starting from the measurement of income levels for countries in 1970 and working backwards with estimates of real national output growth it is possible to construct measures of per capita income for various European countries (including Britain) in the nineteenth century measured in 1970 dollars. We have already encountered in Chapter 2 the index-number problems associated with such an exercise, and we know they are large. Nevertheless, this represents the only possibility of comparing the past and the present, and I propose to take advantage of it. In so doing, of course, I am following the lead of writers such as Kuznets (1971) and Bairoch (1976). For those who are bothered by the length of time involved in the comparisons perhaps it is as well to point out that as far as the analysis of pre-1914 experience is concerned the relative income levels between countries as measured in 1970 dollars are not very much different from what they would be measured in 1910 pounds using official exchange rate comparisons.

Table 3.2 contains a summary of the income levels of some pre-World War I European countries. In the table I have included only the most reliable figures, which, of course, are themselves vulnerable to error. Nevertheless, they are surely good enough for the levels of analysis which are to be undertaken and I doubt that subsequent revision of these data will seriously affect any of the main conclusions about development patterns that I shall draw in this chapter.

Table 3.2 immediately gives us a sense of Britain's position during the 'Industrial Revolution' period relative both to other countries and to income levels at which we expect to see structural change occurring. As would be expected, Britain is the highest income-level country throughout the period. Note, too, that growth was quite widespread in nineteenth-century Europe and that by 1910 many countries had comfortably surpassed the income levels Britain had in 1800 or even in 1840. Further information on income levels, including some less reliable estimates for other countries, is given in Crafts (1983b).

Table 3.2, when compared with Table 3.1, leads us to expect that there might be considerable structural change in nineteenth-century Europe. This was so, and an initial summary of the

Table 3.2. Some Evidence on European Per Capita Income Levels Before World War I in 1970 $ US

	1700	1800	1840	1870	1890	1910
Great Britain[a]	333	399	567	904	1,130	1,302
Belgium				738	932	1,110
Denmark			402	563	708	1,050
Germany				579	729	958
France			392	567	668	883
Sweden				351	469	763
Norway				441	548	706
Italy				467	466	548

[a] Income levels are for the United Kingdom in 1870, 1890, 1910, and England and Wales in 1700.

Source: Taken from Crafts (1983b); the countries are defined on the national boundares as in the years listed. The British figures are based on Feinstein (1972), Deane (1968), and from 1830 back the new estimates given in Chapter 2.

experience can be obtained by employing the same kind of approach that Chenery–Syrquin used. In Table 3.3 are estimates of the 'normal' structural change during nineteenth-century European development. The availability of data does not permit the estimate of such a wide range of equations as was possible for Chenery–Syrquin.

The main purpose of Table 3.3 is to enable us to make comparisons between British experience and European experience. Some readers may find it interesting to compare Table 3.3 with Table 3.1, however. This need not concern us unduly, but some points are of note. The nineteenth-century Europeans appear to have had considerably less demographic pressure, especially because death rates were much higher, and they invested rather less at all income levels. It is noteworthy that the Europeans at income levels from $400 to $900 appear to have been able to employ higher proportions of the labour force in industry, but they too, at these income levels, experienced a substantial productivity gap, with agriculture using up a much higher proportion of the labour force than the proportion of income that it generated.

Before we examine Britain's development transition against this background we should briefly consider the diversity of nineteenth-century European experience. Whilst it is true that the goodness of fit of my equations underlying Table 3.3 was very comparable to

Table 3.3. The Development Transition in Nineteenth-Century Europe

	Per Capita Income Levels in 1970 US dollars				
	$300	*$400*	*$550*	*$700*	*$900*
Crude Birth Rate	38.8	36.5	34.0	32.0	30.0
Crude Death Rate	28.9	26.4	23.7	21.6	19.5
Urbanization	13.0	21.3	30.5	37.5	44.7
Percentage of Labour[a] Force in Primary Sector	72.9	64.3	54.6	47.4	39.8
Percentage of Male Labour[a] Force in Agriculture	75.4	66.1	55.9	48.2	40.1
Percentage of Male Labour[a] Force in Industry	10.1	17.0	24.6	30.4	36.4
Percentage of Income in Primary Sector	54.2	46.5	38.0	31.6	24.9
Percentage of Income in Industry	18.1	21.3	24.8	27.5	30.3
Consumption as % of National Expenditure	83.4	81.5	79.4	77.9	76.2
Investment as % of National Expenditure	10.5	12.2	14.2	15.7	17.2
Government Spending as %[b] of National Expenditure	8.0	7.5	7.0	6.7	6.3
Foreign Capital Inflow as %[b] of National Expenditure	0.9	0.5	0.1	−0.1	−0.4
School Enrollment Ratio	0.174	0.262	0.360	0.435	0.512

Notes:

[a] Male labour force figures are more reliable for our period than female figures, and they are also more compatible with the social tables data (Table 2.1) for England and Wales. Note that mining for the male labour force is part of industry not agriculture. Primary sector includes mining.

[b] These results are less reliable in that the equations did not give a good fit.

Source: Variables are defined as in Table 3.1. All estimates are derived from the regressions reported in Crafts (1984c) and they represent predictions for normal European countries of 10 million population. The equations were estimated using a country dummy variable for Britain, for France, and for Russia and the estimates reported here are for the largest possible samples. Restricting estimation to only the most reliable data set did not materially alter the results. Observations are from 1830 to 1910.

the results obtained by Chenery–Syrquin for 1950–70, and thus generalization is permissible for nineteenth-century Europe in the sense that there are recognizable central tendencies in the evidence, nevertheless we can gain considerably from recognizing and exploring the considerable range of experience.

A simple way to get a sense of this range of experience is to consider the data for countries with income levels c. $550 where development can be expected to be well under way and where the number of observations is quite large, and at c. $900 which represents a high income level by nineteenth-century standards. Table 3.4 displays the available data.

Immediately several things come to light. Table 3.3 has already shown that there is a strong tendency for the proportion of the labour force in agriculture to decrease as income levels rise. Table 3.4 is, of course, not inconsistent with that inference, but note that the countries which made it to $900 had quite strikingly different proportions of the labour force in agriculture. The differences are indeed large enough to suggest the validity of O'Brien and Keyder's insight that there were substantially different paths to modernity being followed during the nineteenth century. A further observation on a similar theme is that individual countries exhibit occasional idiosyncracies; for example, France has a very low birth rate, and Denmark with its very productive agriculture has no sectoral productivity gap even at $550. On the other hand there is a systematic tendency for later arrivals at the $550 income level to have a higher share of the labour force in agriculture at that income level (the rank order correlation is 0.83).

A somewhat better feel for this nineteenth-century experience can be obtained by considering also Table 3.5. This indicates that the rapidly expanding international trade of the nineteenth century took place on the basis of countries having very different export specializations. The situation in 1910 does show one of the reasons why many writers (see, for example, Berend and Ranki, 1980) discuss the development experience in terms of 'centre' and 'periphery', but it also reflects different specializations both among the leaders and the later developers. This also shows up at the $550 income level. Here, however, another tendency shows up, namely that it seems that at this income later developers found it harder to achieve a high share of manufacturing exports in their total exports. Unfortunately the data on exports are not all that good and it is not possible to estimate Chenery–Syrquin equations.

Tables 3.4 and 3.5 are suggestive indeed of the importance of comparative advantage in nineteenth-century development. In other words, that structural change might be linked to a country's position *vis-à-vis* international markets. In turn export specializ-

Table 3.4. Nineteenth Century European Economies at $550 and $900

(a) at $550	*Great Britain*[a] *1840*	*Belgium 1850*	*Denmark 1870*	*Germany 1870*	*Netherlands 1860*	*France 1870*	*Austria 1880*
Crude Birth Rate	35.9	30.0	30.3	38.5	31.6	25.9	37.5
Crude Death Rate	22.2	21.2	19.0	27.4	24.8	28.4	29.7
Urbanization	48.3	na	25.2	36.1	na	31.1	na
Percentage of Labour Force in Primary Sector	25.0	48.9	47.8	50.0	37.4	49.3	55.6
Percentage of Male Labour Force in Agriculture	28.6	51.4	48.2	na	41.3	50.6	57.6
Percentage of Male Labour Force in Industry	47.3	34.4	22.5	na	30.1	28.7	26.3
Percentage of Income in Primary Sector	24.9	27.0	49.0	39.9	na	33.5	na
Percentage of Income in Industry	31.5	24.0	20.0	29.7	na	36.0	na
Consumption as a % of National Expenditure	80.4	na	82.0	81.0	na	78.4	na
Investment as a % of National Expenditure	10.5	na	12.0	15.2	na	12.5	na
Government Spending as % of National Expenditure	7.9	na	6.1	4.3	na	7.0	na
Foreign Capital Inflow as % of National Expenditure	−1.2	na	0.1	−0.7	na	−2.1	na
School Enrollment Ratio	na	0.385	na	na	0.406	0.476	0.374

Table 3.4 continued

(a) at $550	*Sweden 1900*	*Norway 1890*	*Hungary 1900*	*Finland 1910*	*Portugal 1910*	*Italy 1910*	*Spain 1910*
Crude Birth Rate	27.0	30.4	40.7	31.7	31.7	33.3	32.7
Crude Death Rate	16.8	18.0	32.5	17.4	19.2	19.9	23.1
Urbanization	21.5	23.7	na	na	na	na	na
Percentage of Labour Force in Primary Sector	53.5	49.6	64.0	69.2	57.4	55.4	56.3
Percentage of Male Labour Force in Agriculture	53.1	56.0	69.3	69.0	61.0	54.2	59.6
Percentage of Male Labour Force in Industry	24.9	24.0	15.4	12.5	21.7	26.5	13.3
Percentage of Income in Primary Sector	27.2	27.2	na	47.0	na	38.2	na
Percentage of Income in Industry	30.1	22.5	na	25.3	na	23.9	na
Consumption as a % of National Expenditure	84.2	80.8	na	na	na	74.3	na
Investment as a % of National Expenditure	12.0	16.8	na	na	na	16.4	na
Government Spending as % of National Expenditure	5.9	4.8	na	na	na	9.5	na
Foreign Capital Inflow as % of National Expenditure	2.2	2.4	na	na	na	0.2	na
School Enrollment Ratio	0.486	0.435	0.377	0.212	na	0.326	0.366

(b) at $900

	Great Britain[a] *1870*	*Belgium 1890*	*Denmark 1900*	*Germany 1900*	*Netherlands 1900*	*Switzerland 1900*	*France 1910*
Crude Birth Rate	35.2	29.0	29.7	35.6	28.6	28.6	21.3
Crude Death Rate	24.2	20.8	16.8	22.1	13.6	19.3	21.9
Urbanization	65.2	na	38.2	56.1	na	na	44.2
Percentage of Labour Force in Primary Sector	15.3	32.1	46.6	39.9	28.4	34.6	41.8
Percentage of Male Labour Force in Agriculture	20.4	32.1	49.1	35.7	34.4	40.2	40.3
Percentage of Male Labour Force in Industry	49.2	41.5	26.8	43.2	35.8	43.2	33.5
Percentage of Income in Primary Sector	18.8	11.0	29.9	32.2	na	na	27.6
Percentage of Income in Industry	33.5	30.0	26.4	36.6	na	na	38.3
Consumption as % of National Expenditure	80.5	na	75.7	69.9	na	na	75.3
Investment as % of National Expenditure	8.5	na	20.8	22.2	na	na	13.6
Government Spending as % of National Expenditure	4.8	na	6.3	6.8	na	na	8.0
Foreign Capital Inflow as %of National Expenditure	−6.2	na	2.8	−1.1	na	na	−3.1
School Enrollment Ratio	0.168	0.342	0.503	0.561	0.502	0.541	0.577

Notes:

[a] The economic characteristics are all for Great Britain *not* for the United Kingdom.

Sources: are all reported in Crafts (1984c).

ation is influenced by *relative* efficiencies and thus by *relative* factor endowments and technological levels. This theme is one of importance for the understanding of British economic history, and we will build upon it in Chapters 7 and 8.

Britain's position in terms of Tables 3.4 and 3.5 is of great interest. In particular, Britain's characteristics at $550 are worthy of note. At that point Britain had a far higher proportion of exports in manufacturing than any other country achieved, by far the highest urbanization level, by far the lowest proportion of the labour force in agriculture, and had no sectoral productivity gap. All this is prior to the move to 'Free Trade' that occurred with the abolition of the Corn Laws. Britain, in fact, appears to be a *very untypical* nineteenth-century developer.

Table 3.5. Export Experience of European Countries prior to World War I

(a) Percentage of Exports by Value in Manufactures[a] in 1910

Great Britain	76.1	Netherlands	20.9	Portugal	7.7
Belgium	37.0	France	59.2	Italy	28.8
Denmark	5.1	Austria–Hungary	64.0	Spain	24.1
Switzerland	75.5	Sweden	24.1	Greece	0.0
Germany	74.3	Norway	10.9	Russia	3.8

(b) Percentage of Exports by Value in Manufactures at $550

Great Britain 1840	90.5	Hungary 1900	36.6
Denmark 1870	3.8	Sweden 1900	24.4
France 1870	53.2	Spain 1900	24.1
Germany 1870	52.5	Italy 1900	28.8
Norway 1890	8.8	Portugal 1900	7.7

Notes:

[a] 'Manufactures' are not easily defined in terms of nineteenth-century trade statistics and there is some scope for error here. The data were taken from BPP (1913) and Kuznets (1967).

III BRITAIN'S DEVELOPMENT TRANSITION

We can further examine the nature of Britain's 'Industrial Revolution' with the aid of Table 3.6. This table looks at the long run. Income levels prior to 1830 are derived from the growth estimates which were developed in Chapter 2. This is quite an important point since on the basis of Deane and Cole's estimates the income levels would be only $193 in 1700, $251 in 1760, and $309 in 1801. Although our focus will for the moment continue to be mainly on the years to 1840, Chapters 7 and 8 will discuss in rather more detail some aspects of the later years.

Table 3.6 is constructed as follows. The 'European Norm' is the *predicted* value for the given income level from the Chenery–Syrquin-type regressions underlying Table 3.3. The interpretation of the table is that in 1700 the Crude Birth Rate was actually 33.1, whilst the European Norm would be 38.0 at an income level of $333.

The main features of the period usually called the 'Industrial Revolution' are slow income growth and rapid structural change starting from an early eighteenth century position which was already relatively industrialized.

The income levels bracketing the period 1760 and 1840 are in the range of $400 and $550. The 'Industrial Revolution' did not then involve an enormous increase in per capita incomes, as we have already seen in Chapter 2. What is perhaps less well known is that the length of time Britain took to move from $400 to $550—almost 80 years—is unique among the successful developers of nineteenth-century Europe. The range of experience goes from 40 years or so for Norway to less than 20 for Sweden, and averages about 30 per cent of the British time where measurement is possible. The background to this feature of British development is low investment and productivity growth, and this is considered further in Chapter 4. The slow rise in living standards is itself the context against which the standard of living debate should be seen, and Chapter 5 takes up this story.

The rapidity of structural change can be seen in the behaviour especially of labour force shares and relative productivity levels. By 1870 Britain had a share of the labour force in agriculture almost as low as a typical $2,300 country of the post World War II period. The decline in the share of the labour force in agriculture between 1760 and 1840 was especially striking relative to other experience at the same income levels. By 1840 agriculture's share in income and the labour force was virtually the same; this is not surprisingly very different from the experience of latecomers to development, but it is also unlike the experience of other early

Table 3.6. Britain's Development Transition

Year	*1700*	*1760*	*1800*	*1840*	*1870*	*1890*	*1910*
Income Level (1970 dollars)	333	399	427	567	904	1,130	1,302
Crude Birth Rate	33.1	33.9	37.7	35.9	35.2	30.2	25.1
European Norm	38.0	36.5	36.0	33.7	30.0	28.2	27.0
Crude Death Rate	26.5	28.7	27.1	22.2	22.9	19.5	13.5
European Norm	28.0	26.4	25.9	23.4	19.4	17.5	16.3
Urbanization	na	na	33.9	48.3	65.2	74.5	78.9
European Norm			23.2	31.4	44.8	51.3	55.4
Percentage of Labour Force in Primary Sector[a]	57.1	49.6	39.9	25.0	20.0	16.3	15.1
European Norm	69.8	64.3	62.3	53.7	39.7	32.9	28.6
Percentage of Male Labour Force in Agriculture	61.2[b]	52.8[b]	40.8[b]	28.6	20.4	14.7	11.5
European Norm	72.0	66.2	64.0	54.9	40.0	32.8	28.3
Percentage of Male Labour Force in Industry	18.5	23.8	29.5	47.3	49.2	51.1	54.3
European Norm	12.6	16.9	18.6	25.3	36.5	41.9	45.2
Percentage of Income in Primary Sector	37.4	37.5	36.1	24.9	18.8	13.4	10.3
European Norm	51.4	46.6	44.8	37.2	24.8	18.9	15.1

Percentage of Income in Industry	20.0	20.0	19.8	31.5	33.5	33.6	31.8
European Norm	19.3	21.3	22.0	25.2	30.3	32.8	34.4
Consumption as % of National Expenditure	92.8	74.4	76.8	80.4	80.5	81.6	73.8
European Norm	82.7	81.5	81.1	79.2	76.2	74.8	73.8
Investment as % of National Expenditure	4.0	6.0	7.9	10.5	8.5	7.3	7.0
European Norm	11.1	12.2	12.6	14.4	17.2	18.6	19.5
Government Spending as % of National Expenditure	4.8	12.7	15.3	7.9	4.8	5.9	8.2
European Norm[c]	7.8	7.5	7.4	7.0	6.3	5.9	5.7
Foreign Capital Inflow as % of National Expenditure	na	na	0.6	−1.2	−6.2	−5.2	−11.0
European Norm[c]			0.5	0.1	−0.4	−0.7	−0.9
School Enrollment Ratio	na	na	na	na	0.168	0.385	0.542
European Norm					0.514	0.582	0.626

Notes:

[a] This includes mining and therefore differs both from Deane and Cole's estimates for agriculture and the discussion of the social tables in Chapter 2.

[b] Corrected from Chapter 2 to allow for the omission of female domestic service. For males mining is included with industry.

[c] The regressions for these variables give a very poor fit.

developers. Thus from Table 3.4 we can see that France in 1910 at around $900 still had a large sectoral productivity gap with the primary sector's share of income at 27.6 per cent and the labour-force share at 41.8 per cent; this last figure, of course, being about the share of the labour force that Britain's economy had had in the late eighteenth century at about $400. This 'early release' of labour by British agriculture has not been adequately dealt with by economic historians and is a main theme of Chapter 6.

The earlier development of the economy also shows through in Table 3.6. It is noteworthy that already in the 1700–60 period Britain had a relatively high proportion of the labour force in industry. Recalling the evidence of Table 2.3 we know that this very much constituted 'traditional' industry, and it is interesting that the proportion of income derived from industry was not particularly high for Britain's income level. Also, of course, Table 3.6 shows a relatively low proportion in agriculture in 1700–60, perhaps in part reflecting the different property rights and institutions operating in Britain compared with the European continent.

Other features of Table 3.6 are worth noting and bearing in mind for the later chapters of the book. In particular, they are to do with Britain's performance in the accumulation of capital. In terms of European comparisons the British experience was one of relatively low rates of accumulation of physical and human capital at home, and relatively high rates of accumulation of capital abroad. In broad outline this is well known to textbook readers (see, for example, Floud and McCloskey, 1981). Nevertheless, Table 3.6 provides quite striking comparisons. British home-investment levels peak in the mid-nineteenth century at an income level in the $500–800 range and then fall off. Perhaps not always realized is that over much of the second half of the nineteenth century consumption was unusually high and that school enrollment was unusually low. The figure given in Table 3.6 for school enrollment is rather unfair to Britain in that it ignores Sunday Schools, which were relatively important, but the basic picture at around 1890 is a good indication of the lower efforts at human capital formation. That Britain's experience differed from other European capital accumulation may, of course, simply reflect different investment opportunities (returns) available either to Britain *qua* Britain or to Britain *qua* industrializing pioneer.

IV BRITAIN'S INDUSTRIAL REVOLUTION

We can now return briefly to the broad concept of the industrial revolution suggested in Chapter 1 and, as noted, adopted by Deane and Mathias. The purpose of this last section is to review this concept in the context of the empirical evidence assembled in the earlier sections of this chapter.

For Mathias the most important feature of the idea of 'Industrial Revolution' is the 'fundamental redeployment of resources away from agriculture' (1983, p. 2), and he goes on to state that 'the characteristic of any country before its industrial revolution and modernization is poverty' (p. 4). Deane offers a more detailed characterization of the notion of 'Industrial Revolution'. Among the features which would be included are the movement of population from rural to urban areas, intensive and extensive use of capital resources, the movement of labour from activities concerned with the production of primary products to the production of manufactures, and the greater market orientation of economic activity (Deane, 1979, p. 1). Obviously this list bears some resemblance to the development transition of Chenery–Syrquin, but Deane does not specify quantitative criteria, she simply requires that the above 'develop together and to a sufficient degree' (1979, p. 2). Deane does go on to discuss the economy of the eighteenth century in quantitative terms, but her conclusions are again slightly vague: 'the British economy of the mid-eighteenth century displayed (though to a limited extent) a number of the features which we now recognize as characteristic of a pre-industrial economy' (1979, p. 18).

Deane and Mathias are surely right on their own terms in seeing the British economy as experiencing an 'Industrial Revolution' between the mid-eighteenth and mid-nineteenth centuries, but it nevertheless seems worthwhile at this point to attempt to pin down at least some of the details a little more precisely.

A consideration of the economy of England and Wales in 1700 (1688) and 1760 modifies some of the judgements made by Deane and Mathias in the light of more recent research and emphasizes that Deane was right in using the phrase 'to a limited extent' in modifying the term 'pre-industrial economy'.

Deane and Mathias both made considerable use of Gregory King to describe early eighteenth-century England and Wales. As a result they argue for a more agricultural labour force than would now seem to be correct. Deane talks of still 60–70 per cent of the labour force in agriculture in 1750 (1979, p. 14), whereas Table 2.1 led us to a figure of 48 per cent. By the same token Deane implies a very rapid structural change in the deployment of the labour force since in Deane and Cole it is stated that in 1801 only 35.9 per cent of the labour force was in agriculture (1962, p. 142). This seems more abrupt a change than the evidence now available warrants.

The data of earlier tables also allow us to say something about how poor the economy was in the first half of the eighteenth century and to clarify some points of difficulty. Thus Deane argues that Britain was rich relative to most other European countries of the eighteenth century, and had an income perhaps three times that of Nigeria in the early 1960s and nearly at Brazil's level for that period (1979, pp. 9–11). Yet, as we saw earlier, Deane and Cole's estimates imply that the England and Wales of 1700 had an income level of $193 1970 US on the level of, say, Burundi or Rwanda according to Kravis *et al.* (1978), that is, among the very poorest developing countries.

The new estimates give an income level for England and Wales in 1700 of $333 1970 US. This would be about 2.3 times Nigeria's income measured in dollars using official exchange rates for 1970, but only about ⅔ that of Brazil measured the same way (Deane and Cole's estimates would give 1.3 times Nigeria's level). Similarly, the new figures are much more consistent with Deane's argument about Britain's standing in eighteenth-century Europe than are Deane and Cole's (see Table 3.2).

Nevertheless, it is misleading to think of England and Wales in the early eighteenth century as 2.3 times Nigeria's income level in 1970. Obviously there are large differences in the *composition* of consumption and output and, as this chapter has already stressed in another context, relative income levels make an important difference to international specialization opportunities. The point to be made here is a different one, namely that if comparisons are made with countries like Nigeria measured in 'corrected' income levels for 1970, England and Wales would look slightly inferior in 1700 to Nigeria in 1970 ($333 *cf* $356). It should be borne in mind that the official exchange rate calculation is biased against poor countries like Nigeria and 'correction' for differences in price levels shows that their income levels in 1970 were more favourable than Deane's calculation using official exchange rates assumes (Kravis *et al.*, 1978).

Thus, England and Wales should be thought of as at an income level in 1700 similar to that of, say, Nigeria in 1970; above the poorest African countries of 1970 by some way but well below Brazil with an income of $1,102 in 1970 after correction for price level biases. But, if the income level was of only $333, nevertheless the structure of the economy was more industrial than that of either the representative $300 country of the 1950s and 1960s or of nineteenth-century Europe; as a comparison of Tables 3.1 and 3.6 shows, England and Wales in 1700 had more than twice the percentage employed in industry of the former, and nearly one and a half times that of the latter. Moreover, as the social tables in Table 2.1 show and as Deane and Mathias both note, even in 1688 the numbers of those involved in non-market, subsistence production must have been very small. Assuming that we describe all 'cottagers and paupers' as falling into this subsistence category the total would only be 3.7 per cent of the population.

A final point to note about the economy of early eighteenth-century England and Wales is that consumption was a relatively high and investment a relatively low proportion of national expenditure—on Deane's terms this might well be an important part of the pre-industrial-revolution nature of the economy. Levels of capital formation at around 4 per cent of national income are less than a third of the proportion shown for $300 countries in Table 3.1. The low levels of capital formation were, of course, co-existent with relatively low population pressure.

Even if we should stress that pre-industrial is a term to be used with caution about the early eighteenth-century economy, we must nevertheless recognize the very substantial structural changes which had taken place by the mid-nineteenth century. Table 3.6 gives evidence on some of Deane's criteria, and it shows by 1840 already an urbanization level of 48.3 per cent, only 25.0 per cent of the labour force in the primary sector, no sectoral productivity gap, and investment as a fraction of national expenditure 2½ times higher than in 1700. As section III pointed out, although income per head grew only slowly, structural change was rapid.

Thus the economy had a fraction of the labour force in agriculture similar to the early 1950s in France and Germany, and had an urbanization level which would be 'normal' on the basis of Table 3.3 for an economy with income levels similar to France and Germany in the early twentieth century. Moreover, as Table 3.5 showed, exports were almost entirely of manufactures. In terms of urbanization and labour force deployment, as Table 3.1 shows, the economy was reminiscent of a $900 plus economy of the post World War II period. The development transition was not complete, but it was well advanced—in particular, from 1840 on, as Table 3.6 shows, the proportion of the male labour force in industry rose only very slowly to 54.3 per cent in 1910. By contrast the change from 1760 to 1840 was from 23.8 per cent to 47.3 per cent.

It seems possible then to justify the use of the term 'Industrial Revolution' easily enough in terms of this broad concept, particularly if labour force deployment is given the most weight—as in Mathias's definition. Again, in our view of the mid-nineteenth century, however, we must be wary not to exaggerate the degree of completeness of the transformation. The following cautionary points are worth remembering, especially if more weight were placed on factors other than the deployment of labour.

First, the unit of observation in 1840 is *Great Britain*. At that time (before the famine) Ireland had a population which was much more agricultural than that of Britain and was an important part of the United Kingdom's population—in 1841 the working population of Ireland was 47 per cent that of Britain and it was more than half agricultural (Deane and Cole, 1962, p. 146). By no means would a United Kingdom analysis alter the fact that the structural change in the use of labour of 1760–1840 was pronounced, but the extent of the change would be rather less dramatic.

Second, although the economy was a market economy it was not completely a national market. Although markets for goods were generally well integrated as far as measurement can be made (Crafts, 1982), there were exceptions. Thus, coal prices varied dramatically within England—the price in the dearest counties being five times that in the cheapest in 1843 (BPP, 1843). Moreover, the market for unskilled labour seems not to have been completely unified; in 1833 real agricultural wages in the North Midlands, for example, were 40 per cent higher than in the West of England (Crafts, 1982, p. 66).

Third, the economy had obviously become more capital intensive. Feinstein's figures show Britain's gross stock of reproducible industrial capital to be £36 million in 1760, £277 million in 1830, and £697 million in 1860 at 1851–60 constant replacement cost (1978, p. 42). Together with the evidence earlier given on the composition of the labour force, this suggests capital per industrial worker rose from £48.0 in 1760 to £92.3 in 1830 and £148.3 in 1860. Nevertheless, the investment fraction of national expenditure did not rise particularly high, even during the mid-nineteenth-century railway boom, by comparison with other European economies (Table 3.6).

Fourth, Britain consumed far more coal per capita than any other European country. In 1850, consumption was 2.29 tons/person, a figure not reached by Germany until 1880 (Mitchell, 1975). Yet in 1869 46.6 per cent of industrial uses of coal were in iron manufactures and cotton textiles (BPP, 1871).

Fifth, although British exports were almost entirely of manufactures, textiles accounted for 65.3 per cent of their value (Davis, 1979, p. 101). Indeed, across the whole range of industry labour productivity in Britain was not enormously high (see Chapters 4 and 6).

Sixth, by no means was Britain a country composed of large-scale industrial corporate enterprise; Musson points out that in 1851 'out of the total number of masters (129,002) who made returns, more than half (66,497) employed 5 men or less, while a further third (41,732) employed no men or did not state the number employed' (1978, p. 68).

Thus, while we can say that a fundamental redeployment of resources had taken place by the 1840s when Britain had an employment structure quite unlike the rest of Europe, it must be remembered that much of the industrial activity of the economy remained small-scale, little affected by the use of steam power and characterized neither by high productivity nor comparative advantage.

4
Sources of Economic Growth

I INTRODUCTION

THIS chapter is concerned with the changes in the supply side of the economy which raised productive potential and thus permitted economic growth. As such we will be examining the proximate sources of economic growth. In doing so the normal economists' distinction between growth resulting from the use of additional factors of production and from extra output per unit of total input will be used. In other words, we shall be engaged in 'growth accounting' which is assessing after the fact where growth came from.

It must immediately be recognized that this is at once a tricky and a rather limited endeavour. It is tricky because there are important difficulties of measurement and also conceptual problems in imputing growth to various sources. In particular, it is important to remember that the quality as well as the quantity of factors of production may be thought to change over time. it is a limited exercise because of the restricted level of explanation which can be obtained. Thus, at best, we can answer questions of the kind, 'how much extra growth came from increases in capital formation?' rather than questions like 'why did capital formation and (hence) economic growth increase?' This level of explanation is, of course, in keeping with this book's declared aim of seeking to describe the mechanisms of economic growth, but is obviously much less ambitious than trying to find the origins of the industrial revolution.

The chapter does seek in section IV to disaggregate productivity growth somewhat by sector, and the results are interesting as they further develop our ideas on the areas of the economy which were being transformed during the industrial revolution. Discussion of the mechanisms underlying the transfer of resources between sectors, notably between agriculture and manufacturing, is deferred until Chapter 6, however.

The importance of increases in investment relative to increases in productivity as sources of economic growth has been much discussed in the post-war literature on the industrial revolution. The issue is quite well aired in Crouzet (1972), who points out that by that date it was becoming conventional wisdom to doubt the importance for the industrial revolution of 'a notable acceleration in capital accumulation' (1972, p. 19), although he notes that '[the] idea that the capital formation proportion doubled during the industrial revolution, first put forward by Lewis, was elaborated and made widely known by W. W. Rostow' (1972, p. 11). Although cautious, Crouzet seems to accept that productivity growth was considerably more important than capital formation in raising the rate of economic growth after 1780.

Since Crouzet wrote the picture has changed somewhat. Whilst Crouzet based his arguments on the work of Deane on capital formation (1961), he did not formally present any results in a growth-accounting framework. Subsequently, Feinstein (1978) has presented improved estimates of capital formation and has explicitly made estimates of the sources of economic growth. In doing so Feinstein, of course, used Deane and Cole's estimates of output growth and this chapter will correct Feinstein's results to take account of the new estimates of Chapter 2. Feinstein gave higher estimates for the share of national expenditure devoted to investment than had Deane, and combining these with Deane and Cole's national income estimates has argued that 'contrary to the view tentatively advanced by Deane and Cole and now widely (and sometimes dogmatically) accepted, the investment ratio did rise during the eighteenth century, and by quite a substantial margin' (1978, p. 90). In his calculations Feinstein found that 62.5 per cent of the increase in growth in 1801–30 over 1761–1800 was attributable to total factor-productivity growth and 22 per cent to faster growth of the capital stock (1978, p. 86).

Nevertheless, although Feinstein's estimates have redressed the balance a little, it remains true that the literature stresses the contribution of widespread productivity growth to economic growth during the industrial revolution period. McCloskey's view is summed up by the phrase 'ingenuity rather than abstention governed the industrial revolution' (1981a, p. 108), whilst von Tunzelmann adds to this by emphasizing the contribution of many small improvements in productivity: 'It is the variety and pervasiveness of innovation during British industrialization—concentrated in some sectors more than others it is true—which will be argued as having been the key to the growth process' (1981, p. 143).

The new estimates of Chapter 2 suggest that these views are in need of substantial revision. The changed pattern of growth exhibited in Table 2.11 implies that both the rate of capital formation as a share of national expenditure and the rate of productivity growth are different from the figures given by Feinstein. In fact, the lower growth of national output in the new estimates implies lower productivity growth. In turn a lower calculated rate of productivity growth indicates that innovation may have been rather less pervasive in the non-revolutionized sectors of the economy than writers like von Tunzelmann have assumed.

II INVESTMENT

The estimates for capital formation presented by Feinstein (1978) represent an important advance in knowledge, and I have made no attempt to improve on them. This section seeks to put Feinstein's investment estimates in the context of the new estimates of economic growth presented in Chapter 2 above. In calculating gross domestic investment as a share of gross national expenditure Feinstein used his constant price estimates for capital formation, and calculated national income in constant prices working backwards from 1851–60 using Deane and Cole's estimates of growth. Chapter 2 has argued that Deane and Cole's estimates of growth are too high, which would imply that Feinstein's estimates of national expenditure, particularly in earlier years, are too low, thus causing his estimates of investment as a proportion of national expenditure to be too high.

Table 4.1 gives revised estimates of investment's share in national expenditure. Some of the figures have already been shown in Table 3.6. The figures are given for the current price ratio as would be the normal practice in the Chenery–Syrquin approach. In fact the figures for 1801–31 are derived using Deane and Cole's figures for current incomes which are not revised by Chapter 2, and thus the results for that period are exactly what Feinstein would have obtained had he chosen to present estimates of the investment ratio measured in current prices.

Table 4.1. Gross Domestic Investment as a Proportion of Gross National Product (%)

New Estimates		*Feinstein*	
1700	4.0	1761–70	8
1760	6.0	1771–80	9
1780	7.0	1781–90	12
1801	7.9	1791–1800	13
1811	8.5	1801–10	11
1821	11.2	1811–20	11
1831	11.7	1821–30	12

Sources: Feinstein; from Feinstein (1978, p. 91). New Estimates; Figures for gross domestic capital formation plus stockbuilding in current prices from Feinstein (1978, p. 41, 69) divided by national income estimates in current prices from Deane and Cole (1962, p. 166) for 1801–31. For 1760 the national income figure is 232/200 × Lindert and Williamson's figure of £66.4m for England and Wales's national income of 1759. For 1780 the national income figure is derived using the new estimates for real product growth and adjusting for inflation according to the Schumpeter–Gilboy price index, (Mitchell and Deane, 1962, p. 469). For 1700 the estimate is derived using the argument that the capital to output ratio was constant from 1700 to 1760 which is consistent with Deane's assessment (1973, p. 355) and then employing the relation that the growth of the capital stock is the gross investment ratio/the gross capital to output ratio. This was calculated as 5.8 for 1760 using the new estimates for growth and Feinstein's capital stock estimates (1978, p. 84, 86). The capital stock was assumed to grow at 0.69 per cent per year for 1700–60 so $I/Y = 5.8 \times 0.69 = 4.0\%$.

The new estimates given in Table 4.1 do much to restore the picture of gradual advance in the share of expenditure devoted to investment advanced by Deane and Cole (1962, pp. 263–4). The advance in the investment ratio does not match the expectations of Rostow and Lewis if the 'Industrial Revolution' (or 'Take-Off') is seen as a twenty year period at the end of the eighteenth century. However, a doubling of the investment rate between 1760 and 1830 is shown by these estimates, and moreover the investment share of expenditure of 11.7 per cent in 1831 is almost three times the figure for 1700. The new estimates, of course, show a less rapid rise in the investment rate than do Feinstein's figures. They also do not repeat Feinstein's surprising result that investment as a share of GDP peaks at the time of the start of the wars with France.

Chapter 3 made it clear that the investment rates shown in Table 4.1 were low compared with the norms for countries at income levels similar to Britain in this period. It should be noted, however, that this does *not* indicate that personal consumption was high early on. On the contrary, during 1760–1840 consumption was rather comparable with the European norm. During much of this period government spending was relatively high because the country was engaged in war, and by the 1830s (as Table 3.6 shows) foreign investment was starting to use up some savings. It is indeed a plausible conjecture that pursuit of war reduced capital formation and that the crowding-out of private investment by military spending is partly responsible for the slow rise in the investment rate, and perhaps for the hesitant increase in productivity growth prior to the 1820s.

Economic historians have for a long while been accustomed to the argument that the British economy during the Industrial Revolution had a substantial savings potential. In 1800, as Table 3.6 shows, personal consumption was only 77.4 per cent of national product. That the economy could withhold so much from personal consumption indicates that income was on average well above subsistence. Moreover, during the 'Industrial Revolution' the top 10 per cent of income receivers had around 45 per cent of incomes (Lindert and Williamson, 1983a, p. 98) and thus, in principle, had a considerable ability to accumulate assets. The recent literature has therefore tended to stress the importance of mobilizing savings by the development of an improved financial system (Crouzet, 1972, p. 69) as the reason for more investment together with more demand for funds for investment as attractive new opportunities arose in connection with new technology (Deane, 1979, p. 178). These arguments are no doubt correct, although they have not been subjected to formal hypothesis testing.

Nevertheless, looking at the economy at the macroeconomic level suggests some further points of interest with regard to saving behaviour during the 'Industrial Revolution'. The first point comes from the theory of the consumption function, and suggests that the increases in saving which occurred during the eighteenth and early nineteenth centuries may to a substantial extent have been a *result* of increased economic growth. The argument can be sketched out very simply as follows.

Suppose in any period consumption depends on permanent income, and let permanent income be a weighted average of last period's permanent income and current income. We have

$$Y_{pt} = (1 - \lambda)\, Y_t + \lambda Y_{pt-1}\ (0 < \lambda < 1) \tag{4.1}$$

$$C_t = K\, Y_{pt} \tag{4.2}$$

$$C_t - 1 = K\, Y_{pt-1} \tag{4.3}$$

But then we have that

$$Y_{pt} = (1 - \lambda)\, Y_t + \frac{\lambda}{K}\, C_{t-1} \tag{4.4}$$

and therefore

$$C_t = K(1 - \lambda)\, Y_t + \lambda\, C_{t-1} \tag{4.5}$$

This would indeed be a very standard textbook formulation of a popular consumption function hypothesis, which will probably work quite well except in periods of substantial inflation. But we can also manipulate (4.5) to show that C/Y will be lower, and thus the savings rate will be higher the faster income is growing. Divide both sides of (4.5) by Y_t to obtain

$$\frac{C_t}{Y_t} = K(1 - \lambda) + \frac{\lambda C_{t-1}}{Y_t} \tag{4.6}$$

and multiply the last term top and bottom by Y_{t-1} to get

$$\frac{C_t}{Y_t} = K(1 - \lambda) + \frac{\lambda C_{t-1} \cdot Y_{t-1}}{Y_{t-1}\, Y_t} \tag{4.7}$$

Note that the higher Y_t is relative to Y_{t-1} the lower will tend to be C_t/Y_t. Of course, the faster is growth, the higher Y_t is relative to Y_{t-1}.

For illustration, we can experiment with values for the components of equation (4.7) to indicate the difference between no growth (an exaggeration of the seventeenth-century situation) and $Y_t = 1.03Y_{t-1}$, the situation around 1830. Reasonable values for K and λ might be 0.9 and 0.7 respectively. In that case the value of C/Y would change with a rise from 0 to 3 per cent growth to yield, after 10 periods, a decrease from 0.9 to 0.844.

The example is, of course, only suggestive and it is not possible to estimate consumption functions for late eighteenth-century Britain. The point of the illustration is simply to argue that a move to higher growth might of itself account for a substantial part of the increase in the savings rate between 1700 and the 1820s, and that we should perhaps be cautious in attributing a large part of the rise to improvements in financial institutions.

The second point comes from consideration of the relationship between investment and population pressure. Faster population growth means that extra investment has to be undertaken in order to maintain the capital-to-labour ratio of the economy; that is, to keep constant the average amount of capital available to accompany the work of each member of the labour force. Failing to maintain the capital to labour ratio can be expected to put downward pressure on income per head—a point which is elaborated further in Section III. Actually, the economy found it hard to maintain the growth of capital at a rate in excess of the labour force during the period 1760–1830, as Feinstein's figures show (1978, p. 86) and as Table 4.2 also reflects.

If for the moment we assume, as Feinstein does, that population and labour supply grew at the same rate then we can calculate the required increases in the net investment rate to maintain the capital-to-labour ratio in the face of the increased population growth rate. Increases in the population growth rate were implicitly shown in Table 2.11; population growth was at 0.38 per cent per year for 1700–60, 0.69 per cent for 1760–80, 0.97 per cent for 1780–1801, and 1.45 per cent for 1801–31. Thus to maintain the capital-to-labour ratio increases in the growth of the capital stock would have needed to match these population growth increases. The growth rate of the capital stock equals the share of national income devoted to net investment divided by the net capital-to-output ratio, which was about 2.5 in eighteenth-century Britain, and a rise of about one per cent per year in the population growth rate would thus require about an extra 2.5 per cent of national income to be devoted to net investment to compensate, perhaps just under 4 per cent extra given over to gross investment.[1] We immediately see from Table 4.1's new estimates that the rises in the home investment rate obtaining before 1821 were only just sufficient to cope with the extra population growth, but that after the wars with France the investment rate rose to a level giving a margin in excess of the demographic requirement to maintain the capital-to-labour ratio.

[1] The increase in capital ΔK by definition equals net investment, NI. Let the net capital to output ratio, K/Y, be defined as v. The net investment rate $(\mathrm{NI}/Y) \equiv i$. Then the rate of growth of the capital stock, $\Delta K/K \equiv iY/vY$, that is $\equiv i/v$.

This notion of demographic pressure on the investment ratio links up in an interesting way with the recent research on demographic history by Wrigley and Schofield (1981). Wrigley and Schofield have suggested that whilst mortality in pre-industrial England may not have been very sensitive to small changes in income per head, fertility was perhaps more sensitive—in particular, they argue, because a lagged effect of higher real wages would be to make marriage occur at earlier ages, reduce the proportion of the population who never married, and raise illegitimacy (1981, p. 268). Indeed, Wrigley and Schofield's estimates suggest that the crude birth rate rose from a low of 26.8 in 1661 to 35.9 in 1771, and a peak of 41.9 in 1816, before declining to fluctuate between 34.9 and 36.2 between 1831 and 1871 (1981, pp. 528–9).

Wrigley and Schofield's work has been further extended by Lee (1980), who reports an econometric estimate that the elasticity of fertility to the real wage rate was of the order of 0.3. Thus, the economy of the eighteenth century faced a (mild) Malthusian Trap in the form of economic growth triggering off population pressure which threatened the ability to sustain higher living standards. Wrigley and Schofield argue that in previous epochs similar problems had proved too much to cope with, and that income per head as a result did not exhibit sustained advance (1981, p. 440).

The nineteenth-century economy triumphed over the Malthusian Trap. Eventually economic growth provided strong economic incentives to reduce family size as children became 'relatively more expensive' (Crafts, 1984b) and as methods of birth control improved and became more widely diffused. Before this, however, the productive potential of the economy outstripped demographic increase and it is the increase in productive potential which is examined in the next section.

This section has suggested that the industrial revolution down to the end of the wars with France did not see any dramatic rise in the investment rate. The rise in the savings rate between 1760 and the 1820s is much what might be expected from conventional consumption-function theory, and does not indicate any spectacular success in financial institutions' mobilizing of funds as opposed to the effects of rising productive potential. Increases in investment were required to cope with extra population pressure triggered off by economic growth in the late pre-industrial revolution period, and in the period down to the 1820s the contribution of higher investment rates was essentially to prevent the capital-to-labour ratio from *falling* and thus tending adversely to affect income per head.

III SOURCES OF ECONOMIC GROWTH IN THE AGGREGATE

In providing answers to the question 'how much did capital formation and productivity growth respectively contribute to economic growth?' the analytical tool used by Feinstein and McCloskey in the papers cited in Section I above was the 'sources of growth accounting framework'. This approach has indeed been very widely used by economists seeking to understand the process of economic growth; for example Denison (1967) and Matthews *et al.* (1982). The technique provides a way of imputing growth to different kinds of supply-side change and offers a quantitative answer to the question posed by Crouzet concerning the importance of faster capital accumulation.

The approach regards the growth rate of output as the outcome of the growth rate of inputs on the one hand and the growth rate of productivity of those inputs on the other. The growth rate of factor inputs is measured in terms of a weighted average of the growth rates of the factors of production—in this section we consider these as capital, labour, and land, but in principle further disaggregation could be undertaken when data permits. The weights are intended to reflect the 'importance' of the particular input in the productive process, and the sources of growth approach assumes that appropriate weights are found by measuring the input's share in costs: for labour the share of wages in national income, for capital the share of profits, and for land the share of rents.[2] Total factor-productivity growth is then measured as a residual, that is, the difference between the rate of growth of output and the weighted average growth of total input. We have

$$\frac{\Delta Y}{Y} = \alpha.\frac{\Delta K}{K} + \beta.\frac{\Delta L}{L} + \gamma.\frac{\Delta T}{T} + r^* \qquad (4.8)$$

[2] Again we confront an index number problem in the measurement of total factor input. The weights can be justified theoretically in a neoclassical, constant returns to scale, competitive economy producing a single good with homogeneous factors of production where the share of profits will represent the elasticity of output with respect to change in capital, etc.

where $\Delta Y/Y$ is the rate of growth of output, α is the share of profits in national income, $\Delta K/K$ is the rate of growth of the capital stock, β is the share of wages in national income, $\Delta L/L$ is the rate of growth of the labour force, γ is the share of rent in national income, $\Delta T/T$ is the rate of growth of land, and r^* is the rate of growth of total factor productivity. So the contribution of capital-stock growth would be taken to be $\alpha.\Delta K/K$.

Obviously there are various measurement problems to be considered when we seek to use (4.8) for the period of the industrial revolution. We must also be cautious in interpreting the results of (4.8), which is an *accounting identity*. It does not therefore answer deeper questions of causation. For example, in the last section it became clear that the rise in savings and capital formation during the industrial revolution may well have been to a considerable extent a *result* of faster growth. Equation (4.8) does not account for such feedbacks, it deals only with proximate sources of growth. Second, we know that much technological change is in practice 'embodied' in new forms of capital equipment. The residual measures, then, additions to the *quality* of the capital stock amongst other things, since at best our measure of capital inputs will subsume only increases in quantity. The results obtained by applying (4.8) to the crude quantity data on factor inputs in use will thus tend to undervalue the importance of capital formation to the growth process. Third, sophisticated tales about the causes of faster economic growth during the industrial revolution might want to allow for interactions between the growth rates of various factors and productivity. For example, Cole (1973, p. 348) argued that faster population growth from the 1740s on stimulated the growth of incomes per head by raising demand pressure and leading to induced innovations. Again, such claims are beyond the scope of measurement of the simple accounting identity in (4.8).

It must be emphasized that the problems of interpretations described above are not only encountered by users of the sources of growth methodology. They are inherent in all discussions of the contribution of capital formation to economic growth. Setting out a specific formula does, however, have the advantage of making explicit these difficulties. It also offers some scope for comparison with other countries and periods where the same procedure has been used.

In order to apply (4.8) data can be obtained from the work of Feinstein, Deane and Cole, and Wrigley and Schofield. Obviously the data are fallible, as the authors make clear, and as Chapter 2 strove to show. This does impinge particularly on estimates of r^*, the growth of total factor productivity, which emerges as a residual from the use of (4.8) as an identity. That is, values for everything else are put into (4.8) and r^* is then found by arithmetic. Table 4.2 is a 'best guess' and no more than that.

Values for $\Delta K/K$ and $\Delta T/T$ in real terms were estimated by Feinstein. Given that land was an important factor of production, especially in agriculture, I have explicitly included it in estimating the sources of growth, as did McCloskey (1981a, p. 127); Feinstein only distinguished capital and labour in his sources of growth estimates (1978, pp. 140–2). For $\Delta L/L$ I have used the growth in the numbers aged 15–59 in the population, using Wrigley and Schofield's estimates of that age group as a fraction of total population for England multiplied by Britain's population for 1801 onward (1981, p. 529) and using Wrigley and Schofield's figures for England for the eighteenth century. This gives a slightly lower labour-force growth than using population, as Feinstein did, because rising fertility down to 1816 was lowering the proportion aged 15–59.

α, β, and γ are taken to be 0.35, 0.5, and 0.15 respectively. These estimates have to be regarded as somewhat tentative, although they are very similar to McCloskey's figures and are consistent with Feinstein who, in effect, takes $\alpha + \gamma = 0.5$ and $\beta = 0.5$. The difficulty is that, particularly in the eighteenth century, many income receivers were getting 'mixed' incomes; that is, incomes including both 'profits' and 'wages'. Deane and Cole's figures for 1801 show 44 per cent of incomes as employment, a fraction which rises to 49 per cent by 1860 (1962, p. 301). The 44 per cent figure must be somewhat too low, and hence the guess at 0.5 for β. The share of rent does seem to decline after 1830, but I have made no adjustment for that. Note that fairly small changes in the values of α, β, and γ would not radically alter the results of Table 4.2.

The main difference in the results of Table 4.2 as compared with those obtained by earlier writers comes not from changed measurements of total factor input but from the use in calculating the value of r^* of the new estimates for $\Delta Y/Y$ based on the work of

Table 4.2. New Estimates of Total Factor Productivity Growth (% per year)

	$\Delta Y/Y$	Due to $\Delta K/K$	Due to $\Delta L/L$	Due to $\Delta T/T$	TFP growth
1700–60	0.69	0.35 × 0.7	0.5 × 0.3	0.15 × 0.05	0.3
1760–1800	1.01	0.35 × 1.0	0.5 × 0.8	0.15 × 0.2	0.2
1801–31	1.97	0.35 × 1.5	0.5 × 1.4	0.15 × 0.4	0.7
1831–60	2.5	0.35 × 2.0	0.5 × 1.4	0.15 × 0.6	1.0

Source: See text. Note also that $\Delta T/T$ for 1700–60 is a rough estimate based on a comparison of Jones (1981) and Feinstein (1978). 1831–60 is from Feinstein, but with a specific allowance for land. $\Delta K/K$ for 1700–60 is based on Deane's (1973) argument that the capital-to-output ratio was constant in this period.

Table 4.3. Feinstein's Estimates of Total Factor Productivity Growth (% per year)

	$\Delta Y/Y$	Due to $\Delta K/K$	Due to $\Delta L/L$	TFP growth
1761–1800	1.1	0.5 × 1.0	0.5 × 0.8	0.2
1801–30	2.7	0.5 × 1.4	0.5 × 1.4	1.3
1831–60	2.5	0.5 × 2.0	0.5 × 1.4	0.8

Source: Feinstein (1981, pp. 139, 141).

Chapter 2. Obviously, these lower growth estimates operate to reduce the estimated rate of productivity growth. Feinstein's estimates are given in Table 4.3 for comparison.

The main difference of the present results from Feinstein's is obvious, namely that use of the new estimates eliminates the bulge of total factor-productivity growth which Feinstein's estimates gives for 1801–31. The new estimates show instead a steady rise over the last three periods of the table. They also indicate that the first part of the eighteenth century matched the 1760–1800 period in productivity growth, a result which would be surprising indeed to the 'cataclysmic' account of the industrial revolution. The underlying factors in this result will become clear as sectoral performance is considered in Section IV.

For comparison, similar crude estimates[3] for the UK between 1873 and 1913 indicate a productivity growth rate of 0.5 per cent

[3] By 'crude estimates' I mean that no refinements have been attempted to the measurement of factors of production. Matthews *et al.* also give estimates, having adjusted labour-force growth for changes in education, work-intensity, age, etc. which reduce total factor-productivity growth to zero for this period. (1982, p. 211).

(Matthews *et al.* 1982, p. 208). Thus the picture of Britain's nineteenth-century productivity growth given by the new estimates taken together with Matthews *et al.* would show increases to the mid-nineteenth century and lower productivity growth thereafter. The productivity growth rates given by Matthews *et al.* for later periods are 0.7 per cent for 1924–37, 2.0 per cent for 1951–64, and 2.4 per cent for 1964–73 (1982, p. 208).

The estimates of Table 4.2 also suggest that the very heavy emphasis on productivity growth as a source of growth, to be found in McCloskey (1981a), is somewhat misplaced. Taken at face value the estimates of Table 4.2 indicate that only about 30 per cent of the extra growth in 1801–31 compared with 1700–60 came from productivity growth, about 70 per cent from extra factor inputs (20 per cent from capital formation). The 'full triumph' of productivity growth, 'ingenuity rather than abstinence', awaited the second quarter of the nineteenth century.

Actually, however, there are reasons to think that even in its own terms Table 4.2 exaggerates the importance of 'ingenuity'. In particular, it seems likely that labour-force input growth is undermeasured by the use of what is essentially population growth. Contributors to the standard of living debate have always stressed that rising wages often came from upward drift in the skill composition of the labour force. Williamson (1981, p. 26) gives estimates which suggest that increasing skills might raise the effective growth of labour inputs by 0.1 per cent per year for 1821–61. Also, it has been a common theme of the literature that the industrial revolution saw an increase in work hours per worker per year as changed crop rotations reduced slack winter time in agriculture, and as in industry long factory hours were established and St Monday was diminished. Measurement of this supposition has never been adequately accomplished but Freudenberger and Cummins (1976) can be interpreted to suggest that an addition of around 0.2 per cent per year to labour force growth between 1750 and 1820. All in all, then, it seems that 'ingenuity' (total factor-productivity growth) is a bit overstated and the contribution of labour-force growth a little understated by the measures in Table 4.2.

We can also link Table 4.2 to the discussion of Section II which pointed to the pressure of population growth on investment and the capital-to-labour ratio. Table 4.2 indicates that the excess of growth in capital over growth in labour was small down to 1830. It also emerges that the picture given in Section II was slightly too simplified. We see that the land–labour ratio was falling as land growth failed to keep up with population growth. Given the pressure that this added to the situation, as a matter of arithmetic it is seen that had total factor-productivity growth been zero the capital growth (and investment rates) of the eighteenth century would not have been sufficient to lead to increases in income per head. But note that until the nineteenth century productivity growth was not fast, and Table 4.2 suggests that it was not until 1831–60 that the supply side was capable of giving growth in income per head at the relatively rapid rate of one per cent per year. This is an important point for the understanding of the standard of living controversy which will be taken up in Chapter 5.

IV SECTORAL PRODUCTIVITY GROWTH RATES

Obviously, if data permits, the sources of growth methodology can be applied also to agriculture, manufacturing, or whatever other sub-sector of national product is of interest. McCloskey did attempt a sectoral breakdown for 1780–1860, although he used a variation of the methodology described in the last section (1981a, p. 114). From his analysis McCloskey concluded that the contribution of the unrevolutionized sectors of the economy to productivity change was greater than that of the modernized sectors (1981a, p. 115); although productivity growth per year was lower in the non-revolutionized sectors their weight in the economy was nevertheless so large that McCloskey's calculation credited them with about 56 per cent of the productivity change between 1780 and 1860. Von Tunzelmann argued in support of McCloskey that 'the variety and pervasiveness of innovation . . . [was] the key to the growth process' (1981, p. 143).

The new estimates of growth indicate that these opinions need some restatement. As a first step we can consider productivity growth in agriculture and in manufacturing. For agriculture, Feinstein gives estimates of land and capital growth, and Deane and Cole (1962, p. 143) and Chapter 2 can be used to obtain labour-force growth for 1760–1860. Feinstein and Deane and Cole's data suggest that rent had a share of about 40 per cent, as did wages, with capital taking 20 per cent. The resulting estimates for total factor-productivity growth in agriculture would be 0.2 per cent for 1760–1800, 0.9 per cent for 1801–31, and 1.0 per cent for 1831–60. This underlies an important feature of the structural transformation, namely agriculture's rapid relative contraction in its labour-force share, and indicates that this went along with increasing agricultural efficiency.

We can also have a reasonable guess at productivity growth in agriculture during 1700–60. Chapter 2 indicated that labour-force growth was negligible, and Jones's (1981) discussion suggests land inputs grew hardly at all. We also know from Hueckel (1981) that it was only during the Napoleonic Wars that agriculture became 'capital intensive'. It seems likely then that total factor-productivity growth in agriculture during 1700–60 was not all that much less than output growth, and very probably higher than during 1760–1800. It is this behaviour of productivity growth in agriculture which explains in considerable part the failure of overall productivity growth to rise during the 'early Industrial Revolution' of the late eighteenth century.

Calculations for manufacturing have to be based on capital and labour only as Feinstein's data do not enable the measurement of land used in this sector. Again we have labour-force growth measures from Chapter 2 and Deane and Cole, and capital-stock growth from Feinstein. Taking the share of labour to be 50 per cent in this sector, again based on Deane and Cole and Feinstein, would indicate total factor-productivity growth at around 0.2 per cent for 1760–1800, 0.3 per cent for 1800–30, and 0.8 per cent for 1830–60. Productivity growth in manufacturing was, in other words, probably slow until 1830. It should be pointed out that estimates of capital-stock growth in manufacturing are particularly unreliable, as Feinstein makes clear,[4] and that this conclusion may need eventually to be revised.

In the light of the discussions of Chapters 2 and 3, however, the tentative conclusion that productivity growth in manufacturing was low until after 1830 is not enormously surprising. I have stressed that much of manufacturing remained traditional, that

[4] I have reduced the growth of industrial capital estimated by Feinstein for 1760–1800. Feinstein essentially took capital in this period to grow with Hoffmann's industrial-output index. I have assumed that Feinstein's assumption of a constant capital-output ratio is correct, but have assumed that capital grew at the same rate as the Divisia industrial output-growth index of Chapter 2; that is, at 1.81 per cent per year for 1760–1800.

cotton was untypical, and that the revolutionized sectors were a small part of manufacturing, particularly initially. In fact, a very speculative calculation indicates that cotton might have accounted for around half of all productivity change in manufacturing.[5] Whilst the details of this speculative calculation are, of course, disputable, nevertheless the general message that a few sectors dominated productivity growth in manufacturing and that the general experience was of very slow productivity growth is plausible.

This can be further demonstrated by looking now at how McCloskey's own calculations are affected by the revisions to the growth estimates resulting from Chapter 2. McCloskey based his productivity calculations on *gross product* rather than final product (national income). This is perfectly legitimate as a procedure, of course, and, particularly when sectoral estimates are concerned, may be revealing since individual sectors may well diverge considerably in their success in economizing on intermediate inputs. Moreover, McCloskey (for data reasons) used a comparison between prices of outputs and weighted averages of prices of inputs to measure productivity growth, a procedure which in principle should be consistent with the methodology I used earlier but which in the presence of data problems may not always give the same results. Having measured sectoral-productivity growth McCloskey then weighted the results by the sector's share in gross output to obtain the results reported in Table 4.4.[6]

However, McCloskey's estimate of overall productivity growth comes from Feinstein and Deane and Cole, and the rate of growth of productivity and contribution to productivity change for the 'all other sectors' is derived arithmetically. That is, the 0.55 per cent

[5] Data given in von Tunzelmann (1978, p. 182, 239) are available on spindles, a proxy for capital, for 1788 and 1856 and employment in cotton in 1856. It seems unlikely that employment in cotton could have been less than 50,000 in 1788. Using Deane and Cole's figures in cotton consumed to measure output (1962, p. 185, 187), and taking capital and labour each to have 50 per cent shares, would leave an estimate for total factor-productivity growth of 1.65 per cent per year. An average value for cotton's share in final output between 1788 and 1856 might be around 18 per cent, and cotton's contribution to productivity change would then be (0.18×1.65) = 0.3 per cent per year, compared with overall manufacturing productivity growth at about 0.5 per cent per year for the period. Note that the calculations here would be for final output, not gross output, and are therefore not comparable with McCloskey's (1981a, p. 114).

[6] For an algebraic justification of this procedure see Gollop and Jorgenson (1980).

Table 4.4. McCloskey's Calculations of Sectoral Contributions to Overall Productivity Change, 1780–1860

	(1) Rates of Growth of Productivity (% per year)	(2) Weight	(3) Contribution to National Growth of Productivity % per year (= (1) × (2))
Cotton	2.6	0.07	0.18
Worsted	1.8	0.035	0.06
Woollens	0.9	0.035	0.03
Iron	0.9	0.02	0.018
Canals and Railways	1.3	0.07	0.09
Coastal and Foreign Shipping	2.3	0.06	0.14
Weighted Sum of Modernized Sectors	1.8	0.29	0.52
Agriculture	0.45	0.27	0.12
All Other Sectors	0.65	0.85	0.55
Total		1.41	1.19

Source: McCloskey (1981a, p. 114)

figure in col. (3) of row (9) of Table 4.4 is obtained as 1.19 – (0.52 + 0.12) = 0.55.

The new estimates of total factor-productivity growth in Table 4.2 give a quite different result. The 1.19 figure for overall productivity growth is reduced to 0.71 per cent. As a direct corollary, the contribution of 'all other sectors' to productivity growth is reduced from 0.55 to 0.07 per cent per year and the rate of productivity growth in 'all other sectors' is reduced from 0.65 to 0.08 per cent per year.[7].

The general impression to be obtained is that in much of the British economy productivity growth was slow at least until the second quarter of the nineteenth century. Whilst in Chapter 3 a case was made for stressing rapid structural change in the use of the labour force, and that this justified the conventional use of the term 'Industrial Revolution', this chapter has indicated that the term should *not* be taken to imply a widespread, rapid growth of productivity in manufacturing.

[7] In fact this may be too high since my calculations reported above indicate a considerably higher productivity growth in agriculture than McCloskey finds.

V SOME IMPLICATIONS

If the tentative productivity calculations made in this chapter are somewhere near right, then there are some interesting points which follow. The picture drawn above is of an economy in much of which productivity growth was very slow and in which much manufacturing remained traditional and rather little affected by productivity change. Not only was the triumph of ingenuity slow to come to full fruition, but also it does not seem appropriate to regard innovativeness as pervasive.

If the economy is correctly seen as *not* being pervasively innovative during the industrial revolution, then there is further support for the argument I made in an earlier paper (1977) that it is not clear that it was inevitable that Britain would make the decisive innovations leading to early primacy in cotton textiles (and thus 'revolutionized industry') before, say, France.

Further, if the economy was not pervasively innovative during the early nineteenth century, and the period of rapid productivity growth was a relatively brief mid-nineteenth-century phenomenon, then Britain's relative decline in the face of growing foreign competition after 1870 may also be rather less puzzling than otherwise. It has often seemed to be a problem, for example, to those sensing an entrepreneurial failure in late nineteenth-century Britain to explain quite why the quality of entrepreneurship, as reflected (according to them) by productivity growth, should have declined so markedly (Payne, 1974). The finding that productivity growth was also slow till 1830 may help to remove an apparent paradox.

There are also connections that can be made in terms of international comparisons. Table 3.6 demonstrated that Britain eliminated its sectoral productivity gap between agriculture and non-agriculture at an unusually low income level, and by 1840. It is perhaps now opportune to point out that in 1840 Britain had 47.3 per cent of the male labour force in industry, compared with a European norm of 25.3 per cent, but only 31.5 per cent of income in industry compared with a European norm of 25.2 per cent. This means that at any given income level industrial labour productivity in Britain appears to have been relatively low. A fuller examination of what underlies this fact is given in Chapters 6 and 7. Suffice it to point out here that this relatively low sectoral labour productivity in industry is prima facie at least not inconsistent with the argument of this chapter that capital accumulation per worker was not rapid nor was productivity growth rapid during the industrial revolution.

The productivity of the industrial labour force can be crudely compared with that of France and Germany. O'Brien and Keyder made comparisons for Britain and France and concluded that France had higher industrial labour productivity until the 1890s (1978, p. 90). I have made comparisons in constant prices over time and the results are shown in Table 4.5.

Table 4.5. Labour Productivity in Industry (in 1905–13 £)

	Britain	*France*	*Germany*
1905–13	90.4	83.3	106.8
1895–1904	88.6	71.8	86.6
1885–94	78.4	68.2	68.9
1875–84	75.2	63.7	59.6
1865–74	68.8	–	46.8
1855–64	56.4	51.1	39.5
1845–54	47.0	–	–
1835–44	41.6	37.0	–

Source: Crafts (1984a), Tables 8 and 10.

It must be stressed that future research may require further revisions to these comparisons. However, as the results stand, the best guess seems to be that Britain had a small edge over France in terms of average labour productivity in industry in the mid-nineteenth century. Whilst this is not as dramatic as O'Brien and Keyder's result that France had a lead, it still substantiates a main point of their argument, namely that Britain's development path was not one which yielded particularly high industrial labour productivity. This chapter has provided a start towards accounting for that outcome.

5

Economic Growth and the Standard of Living

I INTRODUCTION

A MAJOR purpose of seeking to measure economic growth in the past is to compare the consumption standards of previous generations with our own. As the discussion of Chapter 2 emphasized, such comparisons are difficult to make and encounter 'index number problems'. Nevertheless, the results obtained in Chapter 2 can be used in conjunction with further evidence on the components of national expenditure to produce estimates of overall changes in real personal consumption during the industrial revolution. Also, the sectoral growth rates derived in Chapter 2 can be used together with trade data to throw further light on increases in aggregate consumption.

The measurement of changes in consumption takes on a special significance in the context of the British industrial revolution for historiographic reasons. One of the most famous debates in economic history is, of course, that which concerns the standard of living of the working classes during British industrialization. In the more recent phases of the debate evidence on consumption derived from Deane and Cole's work was presented by Williams (1966), and subsequently refined by Feinstein (1981) in the light of his revisions to Deane and Cole's capital formation estimates. Obviously the new estimates of growth presented above in Chapter 2 imply a different rate of growth of consumption from that estimated by either Williams or Feinstein, and these new results are presented in Section II of this chapter.

It is, of course, the case that evidence on the growth of consumption expenditure at the national level is only a small part of the information relevant to the standard of living debate. In no way does this chapter seek to resolve the arguments or to claim that the revised national income estimates are a decisive set of findings. On their own such estimates are unhelpful in at least two crucial aspects: (i) they relate to macroeconomic experience and do not deal with the diverse experience of different regions or groups of workers nor do they distinguish workers' consumption from that of capitalists or rentiers; (ii) they do not reveal anything about changes in the 'quality of life' as distinct from the quantity of consumption. The usefulness of the new estimates in regard to the standard of living debate is more in establishing the *context* of the debate.

The contention of Chapters 2 and 3 was that Britain grew only slowly during her industrial revolution, whilst experiencing pronounced structural changes. Moreover, the economy was shown to have struggled to cope with faster population growth. It is against this background that discussions of, say, trends in real wages for particular groups of workers should be viewed.

The importance of this to the historiography can be illustrated as follows. Feinstein's (1981 p. 136) figures show consumption per head growing at 1.08 per cent between 1781/90 and 1821/30 as compared to Lindert and Williamson's preferred estimate for real wages growth for blue collar workers of 0.71 per cent per year between 1781 and 1827 (1983b, p. 13). Evidence such as this has been interpreted by writers like Perkin to indicate that there was a 'considerable shift in income distribution towards the rich' (1969, p. 136–8). By contrast, the new estimates given in Section II below show growth of consumption per head at only 0.58 per cent per year between 1780 and 1831. Evidently, this might require a rather different interpretation of events. In fact, the comparison just made is in need of further refinement and discussion, as Section III indicates, but nevertheless it serves to demonstrate an important role for national income estimates in the standard of living debate.

The characterization of Britain as an economy whose per capita national output and consumption expenditure was growing only very slowly until the 1820s has a further implication, namely that quality of life considerations can be expected to loom very large in any discussion of living standards at least up to that point. The argument is simply that if there was fairly little change in consumption standards then movements in consumption are much less likely to dominate movements in the quality of life. This point is expanded in Section IV where crude attempts to expand the national accounts concept of consumption into a 'measure of economic welfare' are reviewed.

Given that our discussion of consumption must take full account of connections with the standard of living debate, the remaining paragraphs of this introductory section are devoted to a short and selective summary of some of the major contributions in the controversy. Readers who are well versed in this historiography will be less offended if they skip immediately to Section II; readers who feel in need of further introductory material should consult Taylor (1975).

Controversy in this area has indeed been heated, as these quotations from the recent (post World War Two) phase of the debate will show. Hartwell summarized his optimistic views thus: 'consideration of estimates of national income and wealth, of production indexes, of wage and price series, of consumption trends and of social indexes *all* . . . indicate an *unambiguous* increase in the average standard of life' (1959, p. 248). Thompson bitterly opposes this interpretation: 'The people were subjected simultaneously to an intensification of two intolerable forms of relationship: those of economic exploitation and of political oppression . . . For most working people the crucial experience of the Industrial Revolution was felt in terms of changes in the nature and intensity of exploitation' (1963, pp. 168–9).

More cautious but nevertheless sharply divided views were expressed by Ashton and Hobsbawm. Ashton argued that there were both gainers and losers, but asserted his belief that 'the number of those who were able to share in the benefits of economic progress was larger than the number of those who were shut out from those benefits' (1949, p. 38). Hobsbawm was agnostic on economic change but argued for a wider conception of living standards about which he was pessimistic: 'There . . . is no strong basis for the optimistic view, at any rate for the period *c.* 1790 or 1800 on until the middle 1840s' (1957, p. 81), and 'The sociological case for deterioration is far more powerful' (1963, p. 128).

Inevitably, such an emotional debate has several different strands to it which need to be disentangled. Hartwell and Engerman correctly identify at least three different questions: (i) what happened to workers' real incomes? (ii) were the working classes better off than they would have been in the absence of industrialization? and (iii) was there some set of policies which could have made the working classes better off than they actually were during industrialization? (1975, pp. 190–3). All these are valid and interesting questions, potentially.

Nevertheless, they can only be discussed effectively by being made more precise. In particular, it is important to consider what is the reference group of workers to be examined and in what period. For example, Ashton discussed 1790 to 1830, Hobsbawm 1790 to the early 1840s, and Hartwell 1800 to 1850, and, given that we know some of these years were wartime and that the rate of growth of national product was changing during the period from 1780 to 1850, the experience may also differ depending which years are considered. Similarly, whilst Hartwell talked of 'the working class' Hobsbawm generally talks of the 'labouring poor' who are presumably a subset of the working class. Finally, many writers, for example Gourvish (1972), stress that there were large regional variations in changes in living standards.

Even supposing that we decided to confine ourselves to question (i) above and have settled on the reference group and period, there are still important conceptual complications and differences between writers with regard to the notion of real incomes.

To some extent we have already encountered these difficulties in Chapter 2. The standard economist's approach, and one which implicitly is usually invoked in the standard of living debate, is to work in terms of the utility of a representative consumer and to define real income as the minimum cost at the prices of the base year of attaining a given level of utility. As the discussion of Figure 2.1 shows, relative price changes, which as we saw in Chapter 2 were substantial during the industrial revolution, lead to ambiguities in the measurement of changes in real income.

There are several additional points to be considered here. First, the consumers 'representative' of different subgroups of the working classes may have different utility functions. For example, the tastes of 'pre-industrial' workers may have differed substantially from workers in, say, 1840s Manchester. Second, the usual method of approaching real income measurement for workers has been an indirect one of deflating money wages by a price index number. The weights to be used in constructing such a price index number would then be expected to differ for the various subgroups. Actually, as Lindert and Williamson argue, the price index numbers used by various participants in the debate are in any case hard to justify as appropriate to *any* sub-group (1983b, pp. 8–10). Third, the arguments to be included in the utility function evidently include in most writers' view items other than consumption of goods and leisure; in particular workers are usually taken to have preferences concerning the 'quality of life'. The nature of the utility function of the representative consumer has been supposed to be very different by protagonists in the debate. Thus, the representative consumers envisaged by the following writers are not the same. Whilst Boyson argued that 'The Lancashire workman probably preferred his big coal fires and hot water with air pollution to the scenic views with few coal fires, few hot meals and rare hot water under the domestic system' (1973, p. 77). Inglis differs: 'what matter are the changes in the quality of life which are rarely quantifiable: and the changes in the quality of the life of most town and village labourers in the Industrial Revolution period were demonstrably for the worse' (1971, p. 31). Thompson postulated an interdependent utility function: 'The average working man remained very close to subsistence at a time when he was surrounded by the evidence of the increase in national wealth, much of it transparently the product of his own labour, and passing, by equally transparent means, into the hands of his employers. In psychological terms this felt very much like a decline in standards' (1963, p. 318). This argument appears, however, to have been assumed away by all subsequent writers.

From this short summary it will be obvious that this is a complicated debate, and indeed one which may never be fully resolved. In addition to the considerations already mentioned there are also serious data deficiencies which mean that evidence on prices, wages, expenditure patterns, and income distribution is incomplete. Under such circumstances we may be rightly hesitant in describing changes in real income for the representative, average worker, and may have some difficulty in deciding whether a given group did better or worse than others (O'Brien and Engerman, 1981).

It is not surprising therefore that writers like Hartwell (1959) have turned to national income growth figures to provide a reference point and to give an indication of the possibilities; it is also correct to comment, as Hobsbawm did, that the national income data are at best indirect evidence (1963, p. 122). Nevertheless, such data are surely useful in the ways suggested earlier.

II GROWTH OF PERSONAL CONSUMPTION

From the evidence assembled to construct the national output growth estimates in Chapter 2 together with information on exports, imports, investment expenditure, and government spending on goods and services it is possible to deduce the value of consumption expenditure from the following identity

$$Y + M \equiv C + I + G + X \tag{5.1}$$

where Y is home output, M is imports, C is personal consumption expenditure, I is investment expenditure (public and private), G is government spending on current goods and services, and X is exports. The logic behind the identity is of course that total goods supplied ($Y + M$) must equal total goods bought ($C + I + G + X$).

This procedure was used by Feinstein (1981) with the results shown in Table 5.1. Feinstein summed up the results as follows: 'there was no worthwhile improvement in real consumption of goods and services per head during the first six decades of industrialization . . . the 1820s and beyond did see a quite substantial improvement in the standard of per capita consumption' (1981, pp.136–7).

The first task of this section is to re-estimate growth in C from (5.1) using the new estimates of changes in real output derived in

Table 5.1. Feinstein's Estimates of Real Personal Consumption per Head (at 1851–60 prices, £ per annum)

1761–60	9.6
1771–80	9.3
1781–90	9.5
1791–1800	9.8
1801–10	10.5
1811–20	11.3
1821–30	14.6
1831–40	17.9
1841–50	19.4
1851–60	22.9

Source: Feinstein (1981, p. 136).

Chapter 2. In order to facilitate fuller comparison with Feinstein's results and with the evidence on real wages to be given in Section III, the discussion of consumption has been extended to 1850 using Deane's estimates (1968). The outcome of the revised calculations is shown in Table 5.2.

Table 5.2. New Estimates of Real Per Capita Consumption

	Y[a]	C/Y[b]	C/Person[a]
1851	100	0.783	100
1841	76.3	0.804	88.2
1831	59.2	0.795	76.6
1821	47.8	0.770	69.1
1801	32.9	0.768	63.0
1780	25.3	0.755	57.3
1770	23.6	0.864	67.4
1760	22.1	0.744	57.2
1700	14.7	0.928	59.4

Notes:

[a] Y and C/Person are index numbers for which 1851 = 100.

[b] C/Y is measured in current prices and is derived using equation (5.1) and the following data sources:

G: based on Mitchell and Deane (1962, pp. 390–1 and 396–7) for government expenditure less debt repayments.

X-M: based on Feinstein's estimates of foreign investment (1978, p. 69).

I: based on Feinstein's estimates of domestic investment (1978, p. 91).

Y: based on Deane and Cole (1962, p. 166) for the nineteenth century and 1700, Lindert and Williamson (1982) for 1760 and interpolations based on the real growth figures of Chapter 2 for 1770, 1780, and 1821.

The method of calculation adopted in Table 5.2 has been to use the estimates of growth in national output given in Table 2.11 to deduce index number values for real output in earlier years than 1851. Then estimates of C/Y measured in *current prices* and based on the sources shown in the footnotes to Table 5.2 have been used to calculate index numbers for C in earlier years which were then divided by index numbers for population to obtain the index numbers for per capita consumption shown in Table 5.2.

It should be noted that, in effect, this approach amounts to deflating consumer expenditure in current prices by a national output-price deflator rather than a deflator restricted solely to consumer expenditure. This is unavoidable given the available data, and it does introduce some bias, although probably not a large one. This argument can be supported by the construction of a crude index of quantities consumed using the materials collected for Chapter 2. This gives growth of real consumption at 1.9 per cent per year for 1801–31 (compared with 2.1 per cent implied by Table 5.2) and 0.95 per cent per year for 1760–1801 (1.05 per cent). The details of the calculation are given in the appendix to this chapter, but the intuitive reasons for quantities of consumption growing a little more slowly than output overall can be enumerated fairly easily.

First, agricultural output was a more important part of consumer spending than national output as a whole. Second, cotton goods were a less important part of consumer spending than national output because a very high proportion of cotton goods were exported (Deane and Cole, 1962, p. 185). Third, an important part of consumer spending went on paying for the distribution of goods, domestic and personal services, and rent where productivity and output growth were not especially rapid; together 'trade and transport', 'domestic and personal', and 'housing' were 28.2 per cent of current incomes in 1801 (Deane and Cole, 1962, p. 166). Finally, as Table 3.6 showed, consumer expenditure as a proportion of national expenditure fell markedly from the high point of the early eighteenth century. Partly this was due to war, but a similar trend occurred in peacetime as the discussion of the consumption function in Chapter 4 would have led us to predict. Thus C/Y was 0.864 in 1770 but only 0.770 in 1821.

Table 5.2 shows a similar pattern to that described by Feinstein, although the magnitudes are somewhat different. Between 1770 and 1821 there appears to have been virtually no growth in per capita consumption, whilst after 1821 there appears to be quite considerable growth in this variable. What emerges clearly from the revised estimates is that this result was to a considerable extent the counterpart of an increase in the investment rate (I/Y); Table 4.1 showed that this rose from 6.0 per cent to 11.2 per cent between 1760 and 1821. In addition, the rate of foreign investment ($= (X - M)/Y$) also rose by 1½ percentage points. In an arithmetic sense this accounts for most of the change in C/Y discussed above; in particular years rises in G/Y associated with war also exercised a considerable influence, and this must be borne in mind in comparing particular pairs of years in Table 5.2.

If an increase in the savings rate was one strong influence on consumption, comparing 1770 and 1820 at least, then on the supply side there is a further aspect to consider. In the long run there is some similarity to be observed between the growth of consumption per head and total factor productivity growth; between 1760 and 1801 per capita consumption grew at 0.25 per cent per year and TFP at 0.2 per cent (Table 4.2), between 1801–31 the figures were both 0.7 per cent per year, and in 1831–61 1.6 per cent and 1.0 per cent respectively. In this last period, as was noted in Chapter 4, the capital-to-labour ratio grew significantly and added to consumption standards, but prior to this increases in consumption per head came chiefly from the influence of productivity growth and the capital to labour ratio was hardly rising at all.

Thus this section together with the work of Chapter 4 gives the following context for the discussion of living standards in subsequent sections of this chapter. Overall growth in output per person prior to 1820 was not rapid, and consumption grew somewhat less quickly than output as a whole during 1770–1820. Consumption growth was affected by a rise in the investment (and savings) rate, but until the second quarter of the nineteenth century this was only adequate to maintain the capital-to-labour ratio rather than substantially to increase it, as happened later. Increases in output per person largely depended in this first phase of the industrial revolution on productivity growth, but this was not rapid and only reached one per cent per year in the period 1830–60. The consumption expenditure to total expenditure ratio was also subject to fluctuation as a result of government wartime expenditure, necessitating caution in the choice of period in examining trends in consumption.

III NATIONAL INCOME AND CONSUMPTION ESTIMATES AS A REFERENCE POINT IN THE STANDARD OF LIVING

In this section the evidence of the national income estimates is compared with more disaggregated data on consumption and also with evidence on real wages.

The use of evidence on the consumption of certain goods was proposed by Hobsbawm (1957). He argued that food consumption showed no optimistic tendencies on the basis of admittedly sparse evidence, at least during 1790 to the early 1840s. The most widely used evidence on consumption has been based on data for imports of tea and sugar, although Hobsbawm also cited many extra pieces of evidence.

There are obvious problems with this kind of information. The goods involved were only a very small part of consumer expenditure, prices varied dramatically, and perhaps so did tastes. Furthermore, there are no good estimates available of price and income elasticities of demand for tea and sugar. The data underlying Chapter 2 provide more information than was available to Hobsbawm and, in fact, those estimates tend to confirm Hobsbawm's suppositions about food consumption prior to 1831. During 1760–1800 food consumption was estimated to grow at 0.5 per cent per year, compared with population growth at 0.83 per cent, and, for 1801–31, food consumption grew at 1.5 per cent compared with population growth of 1.45 per cent. It seems probable that per capita food consumption only regained its 1760 level at around 1840.

Nevertheless, both our earlier review of the concept of real income and the measurement of overall consumption growth indicate that to discuss food consumption without taking into account other consumption items is liable to be misleading. Measured consumption growth was indeed slow between 1770 and 1821, but it seems clear that by 1831 real per capita consumption expenditure exceeded eighteenth-century levels by a margin of 10 per cent or more (Table 5.2). Chapter 2 indicated that agricultural goods became relatively more expensive during the period 1760–1820, whilst items like cotton clothing became much cheaper and at no time did workers spend all their budgets on food. As will be noted below, even the poorest agricultural workers generally spent at least a quarter of their incomes on non-food items.

Thus, the national income estimates tend to confirm Hobsbawm's argument that food consumption deteriorated during the early industrial revolution period, but they also suggest that consumption of other items was growing. These items were in part consumed by workers but, of course, the national income figures cannot tell us anything about the distribution of extra consumption among different classes or groups of workers.

An obvious possibility is to seek further insights into the distributional aspects of growth by comparing real wage growth to growth in national output and/or consumption per head. Indeed, this was an approach adopted by Perkin using Deane and Cole's growth estimates and data on real wages derived from data on money wages given by Bowley and Wood and a variety of price index numbers. Perkin concluded that

> the improvement in real wages did not keep pace with real national income per head. Taking the period as a whole from 1790 to 1850, compared with a doubling of the real national income per head, real wages increased by 51 to 73 per cent, according to the price index used . . . Thus on the most 'optimistic' view of the question, there occurred a decisive shift in the distribution of income away from wages. (1969, p. 138).

Since Perkin wrote the data available on both money wages and prices has improved considerably, especially as a result of the work of Lindert and Williamson (1983b). Their 'best guess' estimates for real wages growth for all blue-collar workers show a 38 per cent increase between 1781 and 1827 which *exceeds* the 34 per cent rise that Table 5.2 shows in real personal consumption per head. The effect of Lindert and Williamson's revisions in combination with the consumption estimates of the previous section seems, if anything, to give the opposite indication to that obtained by Perkin. Such a conclusion would be premature, however, since the data on real wages need further exploration, and the explanation of differences between real wage growth and consumption growth is also worth consideration.

A summary of Lindert and Williamson's main results is given in Table 5.3. Although they used a considerably expanded data set their estimates of money wage changes are very similar in fact to those of Bowley (1900) and Deane and Cole (1962). The price index number is chiefly responsible for their revisions to previous estimates of real wages growth and leads them to the following conclusions:

> After prolonged wage stagnation, real wages, measured by the evidence presented here, nearly doubled between 1820 and 1850. This is a far larger increase than even past 'optimists' had announced. It is also large enough to resolve most of the debate over whether real wages improved during the Industrial Revolution. (1983b, pp. 11–12).

The price index number used by Lindert and Williamson is in many ways better suited to the measurement of changes in workers' purchasing power than earlier price indices. In particular, it has weights much more closely based on workers' budgets, has

Table 5.3. A Summary of Lindert and Williamson's Estimates of Real Wage Growth

(a) Money Wages (1851 = 100)

	All Blue Collar	*Farm Labourers*	*Unskilled Non-Farm*[a]
1781	59.64	72.62	54.88
1819	101.84	134.47	99.41
1827	97.59	106.89	98.89
1851	100.00	100.00	100.00

(b) Cost of Living[b] (1850 = 100)

1781	118.8
1819	182.9
1827	140.9
1850	100.0

(c) Real Wages[c]

	All Blue Collar	*Farm Labourers*	*Unskilled Non-Farm*
1781	50.19	61.12	46.19
1819	55.68	73.52	54.35
1827	69.25	75.86	70.18
1851	100.00	100.00	100.00

Source: Lindert and Williamson (1983b, Tables 3, 4, 5).
Notes:
[a] This group of workers is labelled 'Middle Group' by Lindert and Williamson and includes unskilled building workers and miners. It also includes cotton-textile workers, some of whom may be considered 'skilled'.
[b] This is Lindert and Williamson's 'best guess' or 'Southern Urban' index; in their working paper (1980) they also provide three alternative indices for serious consideration as well as twelve inferior ones.
[c] This calculation is made by Lindert and Williamson using the money wages and cost of living indices in the earlier parts of the table.

estimates of changes in rents, and includes cotton clothing, which were omitted from earlier price indices. Those indices due to Silberling (1923), Tucker (1936), Gayer, Rostow, and Schwartz (1953), and Phelps-Brown and Hopkins (1956) were not well suited at all to the task of deflating money wages to obtain real wage estimates (Crafts, 1982, p. 58) and in the Silberling and Gayer, Rostow, and Schwartz cases were in any case intended to be used to study business-cycle fluctuations.

Obviously it is possible to question Lindert and Williamson's price index. No more than any other price index can it claim to be a measure of the 'true cost of living' in the theoretical sense defined earlier in the chapter. Also, it might be argued that many different cost of living indices are required for subsets of the working classes. In a detailed investigation of the standard of living debate such questions would need to be taken up seriously, but at the level of generality of this chapter they are perhaps too pedantic. There are, however, improvements to Lindert and Williamson's price index that need to be made even if all that is sought, as here, is a first approximation of a reference comparison between real wage growth and national-income growth. There are three specific problems; the treatment of clothing, of rent, and of purchases of bread and flour. Of these, the first is by far the most important. Full discussion can be found in Crafts (1985a) but a brief account is as follows.

Lindert and Williamson's approach to clothing prices seriously overestimates falls in clothing prices after 1820 because it is based only on cottons and exaggerates the increase in clothing prices before 1820 by excluding cottons. Their series for rent is taken from evidence for Staffordshire and does not appear accurately to reflect national trends. Finally, the inspection of budgets in Smith (1864) suggests a much bigger weight for bread and a much smaller one for flour than Lindert and Williamson used.

Correction of these problems with Lindert and Williamson's cost of living index leads to the revised estimates given in Table 5.4. The details of the procedures used can be found in Crafts (1985a). It must be understood, however, that data problems are such that no cost of living index can be regarded as very reliable.

Table 5.4. A Revised Cost of Living Index

1780	103.6
1820	135.7
1830	120.5
1850	100

Source: taken from Crafts (1985a, Table 3).

The key feature of the new index, and one which is robust, is that it indicates that the cost of living for workers rose considerably less than Lindert and Williamson's estimate prior to 1820, and fell much less after 1820. In particular, this difference is accounted for by amending their estimates of clothing price changes whilst retaining the broad design of their weighting system, which is indeed much closer to the pattern of workers' budgets as best we know them than were those of the earlier price index numbers.

We are now in a position to make comparisons between real wage growth and national-output and consumption growth. The new estimates based on Chapter 2 and Tables 5.2 and 5.4 are summarized for 1780–1821 and 1821–51 in Table 5.5. Earlier writers' estimates are reported in the second part of Table 5.5. The real wages growth estimates shown include those of Phelps-Brown and Hopkins which are among the most 'pessimistic', and those of Deane and Cole which, especially for 1820–1850, are among the most 'optimistic'.

The main result of the new estimates is that real wage growth and national-output growth appear to be fairly similar in the long run. This is perhaps consistent with the general impression of fairly constant shares of labour and non-labour income which is conveyed by Deane and Cole. Certainly, on these new estimates there would be no strong grounds for asserting that income distribution moved drastically against workers. It remains for us to consider at the end of this section changes in distribution *within* these broad factor shares, a point with which Perkin was much concerned.

The new estimates for real wages growth are quite similar to those of Deane and Cole for the long run, and those of Phelps-Brown and Hopkins for 1820–50. Lindert and Williamson's estimates for real wage growth appear to be very optimistic for 1820–50, and indeed seem very hard to reconcile with national-output growth evidence. In particular, it is difficult to believe that real wage growth should exceed growth of consumption per head and national output per head in 1820–50. We know that working-class savings propensities were very low (Von Tunzelmann, 1982, p. 5) and that the overall savings propensity of the economy changed little (Table 5.2). It seems probable that Lindert and Williamson's estimates for real wage growth in this period are the result of an over-optimistic price index number as was suggested in the discussion of Tables 5.3 and 5.4.

Rather similar reasoning can be applied to 1780–1820. In that period, given that the overall propensity to consume was fairly constant (*C*/*Y* in Table 5.2) and that workers saved virtually nothing, it seems improbable that real wages would grow by much less than consumption per head. In particular, this would seem to

Table 5.5. Comparisons of Real Wage Growth, Per Capita Real Output Growth, and Real Personal Consumption Growth (% per year)

(a) New Estimates[a]

	Real Wages	*National Output per Head*	*Consumption per Head*
1780–1821	0.71	0.42	0.47
1821–1851	0.94	1.19	1.24
1780–1851	0.80	0.75	0.80

(b) Earlier Estimates

	Deane and Cole[b] *Real Wages*	*Phelps–Brown*[c] *and Hopkins Real Wages*	*Lindert–Williamson*[d] *Real Wages*	*Deane and Cole*[e] *National Output/Head*	*Feinstein,*[f] *Consumption/Head*
1780–1820	0.67	−0.03	0.28	0.98	1.08
1820–1850	1.19	0.92	1.92	1.48	1.52
1780–1850	0.93	0.38	1.00	1.19	1.27

Notes:

[a] Based on Tables 5.2 for consumption per head and 2.11 for national output per head. Real Wages use the cost of living index of Table 5.4 and the all blue-collar wages of Table 5.3; real wages growth is estimated for 1781–1819 and 1819–1851.

[b] Based on Deane and Cole's discussion of real wages (1962, pp. 24–5); the money wages figures are those of Bowley-Wood and these were deflated using the Gayer, Rostow, and Schwartz price index.

[c] Obtained using a five year moving average, from Phelps-Brown and Hopkins (1956, pp. 313–14).

[d] Derived from Table 5.3, 'All Blue Collar' workers index using 1781–1819 and 1819–51 for the sub-periods.

[e] Derived from Deane and Cole (1962, p. 78 and p. 282).

[f] Derived from Table 5.1, using 1781/90 and 1821/30 and 1821/30 to 1851/60 for the sub-periods.

argue for the implausibility of the Phelps-Brown and Hopkins measure of real wages and may well suggest the Lindert-Williamson price index is somewhat too pessimistic for 1780–1820.

These arguments actually need refinement. The evidence we have discussed relates to real wage *rates* not earnings. The early 1820s are notorious as a period of very high unemployment in the fact of post-war adjustment and rapidly falling prices. It is quite likely that unemployment levels were above 10 per cent (Lindert and Williamson, 1983b, pp. 14–16), whereas 1780 and 1850 were both times of relatively full employment. Allowing for likely unemployment would perhaps indicate a growth rate of real wage *earnings* of around 0.55 per cent per year for 1780–1820 and 1.2 per cent per year for 1820–50.

In general, then, the argument of this section is that real wage growth was rather similar in the long run to per capita income growth. Living standards did not grow very quickly prior to the 1820s. It seems that the most important underlying reasons are to do with the economy's inability to achieve high rates of growth of productivity or the capital–labour ratio rather than any tendency for real wages growth to be held below the overall rate of economic growth. On this point the evidence does not now seem to support Perkin's contention.

It must be stressed, however, that the discussion thus far has dealt only with the aggregate experience of blue-collar workers. The review of the standard of living debate presented in Section 5.1 above indicated, however, that much of the literature has been concerned with subsets of this group. Perkins has argued that the benefits of economic growth were long delayed for unskilled workers in agriculture and the domestic trades, especially in the south of England, and has claimed that 'It may well be that until the steep decline in handloom weaving and other dying domestic trades after 1840 and the absolute decrease of agricultural labour after 1851, the numbers of those who suffered exceeded those who benefited' (1969, p. 148). Such a conclusion is not, of course, ruled out by the evidence reviewed above.

In general, the data on wages of poorer workers are not good—it may even be that Lindert and Williamson's sample is slightly biased in its coverage towards more fortunate workers. We can get some sense of the validity of Perkin's claim, however, by considering the real wages of agricultural workers. These workers are taken by Lindert and Williamson to be representative of the bottom 40 per cent of all workers (1983b, p. 2) and their earnings have generally been taken to indicate the state of the labour market for the unskilled (Hunt, 1973, p. 4).

Evidence on money and real wages for agricultural workers by region is shown in Table 5.6. It should be noted that the price data used in preparing this table are generally taken from prices of London institutions, and in other cases (cottons and rents) are national estimates. There are good reasons to believe that apart from rent in inner cities regional price variations during the industrial revolution were probably not very large (Crafts, 1982), but further research may eventually lead to proper regional price indices and thus a superior set of estimates. Perhaps more important will be further investigation of regional differences in spending patterns (and thus in the weighting of cost of living indices). In Table 5.6 a crude allowance has been made for differences in regional food-consumption patterns. Table 5.6 is then no more than a useful starting point for future discussions of Perkin's argument.

At first sight this table appears to suggest that Perkin's description is much too pessimistic even for 1824, where only 41.2 per cent of agricultural workers (those in the South Midlands, East, and South West) have real wages lower than in 1780. But unemployment was chronic among agricultural workers in the early 1820s and allowance for this would mean that only the real earnings of agricultural workers in the North could definitely be regarded as higher than in 1780. By 1837 the gains in each of the calculations are sufficient to suggest that the fraction of losers was quite small and did not in general include agricultural workers.

Table 5.6 thus offers some important additional insights into the connection between economic growth in general and workers' living standards. First of all it indicates the need for great caution in assessing the fate of subgroups in the labour force at least prior to the 1830s. Second, it suggests that the pessimists in stressing the experience of groups such as the 'labouring poor' have drawn attention to an important aspect of economic growth, namely that its effects on workers' real wages were uneven. Third, this links quite nicely to the earlier discussions of Chapters 1–4 on the nature of economic growth during the industrial revolution. There it was pointed out that the transformation of the economy was

Table 5.6. Real Wages of Agricultural Workers in Various Regions (1851 = 100)

	1780	*1824*	*1837*	*1851*	*% of agricultural workers in 1841*[a]	*% of all workers in 1841*[a]
(a) Money Wages						
South East	90.8	111.0	117.4	100	16.6	23.3
South Midlands	83.7	96.2	106.7	100	12.0	6.5
East	101.1	113.8	131.9	100	13.2	7.0
South West	88.2	89.9	104.3	100	16.0	11.2
West Midlands	88.0	106.0	115.0	100	13.3	13.4
East Midlands	79.3	101.7	114.0	100	8.7	7.7
North	53.4	94.5	100.0	100	20.2	30.9
(b) Real Wages: deflated using a Southern Agricultural Worker Index[b]						
South East	78.8	80.9	94.1	100		
South Midlands	72.6	70.1	85.5	100		
East	87.7	82.9	105.7	100		
South West	76.5	72.1	83.6	100		
(c) Real Wages: deflated using a Northern Agricultural Worker Index[c]						
West Midlands	68.6	74.8	91.1	100		
East Midlands	61.9	71.7	90.3	100		
North	41.7	66.6	79.2	100		

Notes:

[a] Based on Lee (1979).

[b] The Southern Agricultural Worker index uses the same price data as the general index reported in Table 5.4. It has different weights based on Lindert and Williamson's rural indices modified in the light of Smith's (1864) food-consumption studies. The weights are as follows: rent: 7.3 per cent; fuel: 3.9 per cent; candles: 2.5 per cent; woollens: 8.1 per cent; cottons: 4.1 per cent; shoes: 2.0 per cent; bread: 19.0 per cent; flour: 23.9 per cent; bacon: 10.6 per cent; beef: 1.4 per cent; mutton: 0.7 per cent; potatoes: 6.6 per cent; tea: 4.0 per cent; sugar: 1.5 per cent; butter: 4.4 per cent; food is 72.1 per cent of the weighting.

[c] The Northern Agricultural Worker index is devised in similar fashion. The weighting is as follows: rent: 16.6 per cent; fuel: 3.9 per cent; candles: 2.2 per cent; woollens: 11.7 per cent; cottons: 5.8 per cent; shoes: 2.9 per cent; bread: 3.9 per cent; flour: 33.6 per cent; oatmeal: 1.9 per cent; bacon: 8.1 per cent; beef: 1.0 per cent; mutton: 0.5 per cent; potatoes: 7.8 per cent; tea: 1.7 per cent; sugar: 5.3 per cent; butter: 3.1 per cent. Food is 66.9 per cent of the weighting.

Sources: money wages are from Bowley (1900); the areas are defined as follows: *South East:* Middlesex, Surrey, Kent, Sussex, Hampshire, Berkshire; *South Midlands:* Oxfordshire, Hertfordshire, Buckinghamshire, Northamptonshire, Huntingdonshire, Bedfordshire, Cambridgeshire; *East:* Essex, Suffolk, Norfolk; *South West:* Wiltshire, Dorset, Devon, Cornwall, Somerset; *West Midlands:* Gloucestershire, Herefordshire, Shropshire, Staffordshire, Worcestershire, Warwickshire; *East Midlands:* Leicestershire, Rutland, Lincolnshire, Nottinghamshire, Derbyshire; *North:* Cheshire, Lancashire, Yorkshire, Durham, Northumberland, Cumberland, Westmoreland.

'uneven', with productivity growth concentrated on relatively few sectors which in turn were quite localized. The counties of the South East, South Midlands, East, and South West generally had substantially less than 10 per cent of their labour force in 'revolutionized industry' in 1841 (Table 1.1), whilst the North in all cases is at least 10 per cent and for Lancashire and Yorkshire West Riding is close to 40 per cent. It would seem that productivity growth largely passed by many of the local labour markets of the economy and that the process of migration to the 'North' was unable to spread the gains from productivity growth very evenly in the early stages of the industrial revolution.

It also appears that considerable changes were taking place in the personal size-distribution of income even though the share of wages in national income remained relatively constant. Lindert and Williamson (1983a, p. 98) provide the following evidence based on their revisions of the social tables. They suggest that the share of the top 10 per cent of income receivers in national income rose from 44.4 per cent in 1759 to 45.4 per cent in 1801 and 52.7 per cent in 1867. In money terms this would imply by 1851 that the average member of the top 10 per cent received £125 per year, compared with £5.8 for an agricultural labourer's family member, a ratio of about 21.6 to 1. For comparison, in 1759 the ratio would have been £45 to £4, or about 11.25 to 1. Thus the monetary gap between the top and bottom of society seems to have risen quite sharply in this period, which is no doubt what Thompson had in mind when suggesting the interdependent utility-function argument.

Thus the macroeconomic evidence of growth in both national consumption standards and real wages should not be taken as evidence to settle the standard of living debate, nor should it detract from the importance of examining trends in income inequality and wage differentials. The macroeconomic evidence is certainly not adequate to refute a carefully stated 'pessimist' view of *c.* 1770 to *c.* 1825. Nevertheless, the national output and expenditure data offer useful evidence about distributional trends and are helpful in assessing the plausibility of estimates of trends in real wages.

IV NATIONAL INCOME PER HEAD COMPARED WITH A 'MEASURE OF ECONOMIC WELFARE'

In the standard of living debate many writers place a large weight on changes in the quality of life although economic historians have always found this difficult to deal with. Economists confront similar issues in seeking to measure 'true' economic growth today. Accordingly, attempts have been made to amend national-income accounts estimates of growth. In particular, attempts have been made to impute gains or losses arising from changes in leisure and in urban disamenity (Nordhaus and Tobin, 1972) and changes in life expectancy (Usher, 1980). In their work Nordhaus and Tobin named the resultant calculation a 'Measure of Economic Welfare', of which they proposed several variants. Such work is still relatively new and quite controversial among economists. Recently Williamson (1982 and 1984) has made attempts to apply similar methodology to the standard of living debate. It seems worth considering what this approach has to offer since it potentially allows some grasp of the quality of life problem which, as was noted on Section I, could easily dominate the rather small changes in consumption per head taking place until the 1820s. It will be seen, however, that the assumptions involved are often quite heroic and the empirical results are probably not very robust.

Briefly, the essential nature of the arguments is as follows. Consider again a 'representative consumer', maximizing utility subject to the budget constraint. We can think of, for example, longevity and leisure as goods and imagine that the consumer has preferences and would make trade-offs between consuming goods which are part of GNP and longevity or leisure. Thus, in principle, we can think of willingness to pay for longer life or less work, for each of which there is an implicit price. Consumption of the same goods with higher life expectancy would be a move to a superior position, as would consumption of the same goods with more leisure. Thus, looking only at directions of changes rather than magnitudes, we would expect to find increasing life expectancy tending to add to a 'measure of economic welfare' and decreasing leisure tending to diminish it. Both these effects seem to have been operating during the industrial revolution.

Similarly, we expect that in labour markets which are in equilibrium real wages for unskilled labour will be equal everywhere. The real wage, however, includes non-monetary aspects of work. Thus, work in relatively unpleasant conditions would require a higher money wage to yield the same real wage. We know that during the industrial revolution the proportion of the population whose work required them to live in unpleasant urban surroundings rose. Accordingly, we could expect some part of the increases in nominal earnings accruing to workers to be apparent rather than real gains as they simply represent compensation for increased disamenity.

Some estimates of the magnitudes of the above three effects can be made, although they are not necessarily very reliable. Other aspects of a changing quality of life remain completely intractable, however. For example, Thompson's allegations of feelings of being exploited and oppressed growing among workers may or may not be correct, but are beyond the reach of the 'measure of economic welfare' methodology.

Life expectancy, on average, rose during the industrial revolution. Table 5.7 reports Wrigley and Schofield's estimates for England. Improvements in mortality generate additions to economic welfare; the difficulty is to estimate the value of these additions.

Table 5.7. Life Expectancy at Birth in England during the Industrial Revolution

1701	37.11
1761	34.23
1781	34.72
1801	35.89
1821	39.24
1841	40.28
1851	39.54

Source: Wrigley and Schofield (1981, p. 529)

Usher (1980) uses the following methodology to impute a value for rises in life expectancy. He takes the representative consumer to have the utility function.

$$Y_{(t)} = Y_{(t)}{}^{B} L_{(t)} \tag{5.2}$$

where

$$L = \sum_{j=0}^{N-1} \frac{S_{j+1}}{(1 + r)^{j}} \tag{5.3}$$

S_{j+1} is the probability of surviving one more year at age j and r is the rate of discount. This permits an estimate of the adjusted rate of growth of real income $\Delta\hat{Y}/\hat{Y}$,

$$\frac{\Delta \hat{Y}}{\hat{Y}} = \frac{\Delta Y}{Y} + \frac{\Delta L/L}{B} \tag{5.4}$$

In effect a change in life expectancy is taken to be an exogenous change in the consumer's environment which cannot be bought but for which a price would willingly be paid. Among the problems with this approach is that it is necessary to assume values for r and B for the industrial revolution period, although Usher was able to find ways of estimating those parameters using twentieth-century data. Calculations of $\Delta L/L/B$ are possible for illustrative values of r and B of the order of magnitude estimated by Usher. The results based on Wrigley and Schofield's data would be: for 1780–1821, 0.3 per cent per year for r = 1 per cent, B = 0.45 , and 0.05 per cent per year for r = 5 per cent, B = 0.23 ; for 1821–51 these estimates would be + 0.02 per cent and + 0.03 per cent respectively. Williamson (1984) gives reasons for suggesting that r = 5 per cent, B = 0.23 are quite reasonable guesses for the parameter values.

For 1780–1821 this method of valuing gains in life expectancy would give a big impact in the sense that these gains might be about as large as those coming from consumption of goods. The estimates are based on one particularly strong assumption, however, and one which may mean they are rather too big; the key assumption is that improvements in mortality were *exogenous*. If the gains in life expectancy were purchased via better nutrition then they would be part of the already counted gains in real wages. Recent research by Wrigley and Schofield suggests that the exogeneity assumption may be correct and the earlier sections of this chapter suggested that nutritional standards may well have fallen in this period, but there is a school of thought which asserts that mortality declined only because of nutritional improvements (McKeown, 1976) and the explanation for rising life expectancy remains controversial.

Changes in leisure during the industrial revolution present even greater difficulties. There is some evidence to suggest that the length of the working year for many workers rose after the mid-eighteenth century; Tranter (1981, p. 220–1) suggests that the working year for the average fully employed worker may have risen from 2,500 hours to 3,000 hours between the mid-eighteenth and mid-nineteenth centuries as seasonal underemployment in agriculture fell, factory work replaced domestic work, etc. This estimate must be regarded as very rough, but it is still worth considering what impact this may have had on living standards.

Usher (1980, pp. 135–47) suggests that changes in leisure relative to a base year should be added to real consumption in a given year by valuing them at the real wage of that year. This procedure would then raise the measured living standards of, say, 1760 relative to 1850. In fact, the impact would be quite large: for 1760–1851 it would reduce the rate of real consumption per head growth from the 0.63 per cent per year implied by Table 5.2 to 0.37 per cent per year. Again, there are possible objections to the assumptions involved in such a calculation—did workers' capacity to enjoy leisure change over time? was some leisure time in the mid-eighteenth century worthless? and so on. These points are developed a little further in appendix 2 to this chapter. More important is to note that allowance for changes in leisure would tend to *reduce* the measured growth in living standards, and that the magnitude of the effect *could* be quite big.

Williamson (1982) sought to measure the extent to which gains in real wages were 'spurious' in the sense that they merely compensated workers for accepting reduced levels of amenities in urban areas. In particular, he used infant mortality as a proxy for urban disamenity and estimated regressions to examine how sensitive real wages were to infant mortality, controlling for other variables. Williamson found that the disamenities premium at most amounted to 13 per cent in the early nineteenth-century North of England (1982, p. 234). Since this only applied to a fraction of the labour force, surely no more than 50 per cent in the mid-nineteenth century, this result would imply that corrections to real wages growth on this account would be very small. Williamson's estimates appear quite likely to be disputed (Brown, 1983; Pollard, 1981b) and it remains to be seen whether a case can be made for a much larger disamenities premium.

This section has sought to show some of the limitations of calculations of national-income growth or real wages for the discussion of issues raised in the standard of living debate. A 'measure of economic welfare' is at best only tenuously in the grasp of quantitative economic history, but the possible impact of items not captured by consumption expenditure, particularly of changes in life expectancy and leisure, looms large. Nevertheless, what also emerges is that it would be wrong automatically to assume that all the environmental quality of life considerations are large and negative, and this is a useful, if preliminary, result from the crude attempts at quantification.

V CONCLUDING POINTS

The work of this chapter has established that personal consumption per head grew in real terms during the industrial revolution, particularly after 1820. Between 1770 and 1820 the growth of consumption per head was less than national output per head, and increasing shares of output were taken by investment and government expenditure. The best guess available indicates that in the long run at least growth of real earnings for workers as a whole moved roughly in line with growth of national output per head. Models which postulate a massive shift of income distribution against workers do not seem appropriate.

It has also been shown, however, that whilst useful the above outline does not contribute much towards settling various knotty aspects of the standard of living debate. In particular, in Section III it became apparent that a more disaggregated view suggested many workers may have been losers at least until the 1820s, and that, as a result, great care needs to be taken in defining which workers' living standards and at what time are to be discussed. Indeed, Section III makes it clear that despite better macroeconomic data Ashton's belief that there were more gainers than losers in the working classes in 1830 is still contentious.

Also, in Section IV it was shown that adjustments to the national expenditure evidence on consumption to accommodate the effects of reduced leisure and increased longevity are controversial but could be large. Maybe the effects would cancel out, but not necessarily. Thus the 'quality of life' questions remain unresolved.

Nevertheless, the establishment of the context of slow growth at least until the 1820s and, as Chapter 3 showed, by international comparisons a very slow advance from a $400 to a $550 income level is perhaps of some help. It suggests that those interested in the wider questions relating to alternative economic policies or counterfactual growth paths should consider them against an explanation of Britain's slow growth. Slow rises in living standards can be set against a gradual acceleration of productivity growth and an inability to raise the capital-to-labour ratio until the 1830s. Discussions of how much scope there was for things to turn out differently can take this as their starting point.

APPENDIX 1 TO CHAPTER 5

In section II of the chapter it was noted that an alternative way to try to measure consumption growth would be to construct a quantity index. Such an index is inevitably crude. The one whose growth rates are cited in the text is a divisia index whose weights are approximately the weights in consumer expenditure prevailing in 1801. Some commodities cannot be covered and some are probably over-represented in the index, notably 'trade and transport'. The weights were calculated from Table 2.1, Deane and Cole (1962, pp. 166, 185, 197, 212) and Davis (1979, pp. 96, 114–15). Allowance was made for international trade in agriculture and textiles. The weights listed below were then used to obtain a weighted average of growth in consumption using the data of Chapter 2. It should be noted that the growth of consumption is *not* very sensitive to changes in the weighting, within reasonable limits, and in particular cotton was always a small part of consumer expenditure.

The weights used for the calculation of the consumption quantity index were:

Agriculture	0.407	Leather	0.048
Rent	0.070	Soap	0.005
Cotton	0.034	Candles	0.006
Wool	0.083	Domestic and Personal Services	0.074
Linen	0.015	Trade and Transport	0.233
Silk	0.011		
Linen	0.015		
Beer	0.014		

If the index is recalculated to omit 'trade and transport' entirely then the rate of growth of consumption in each period (1760–1801, 1801–31) is also lowered.

APPENDIX 2 TO CHAPTER 5

Economists have adopted different ways of dealing with the valuation of changes in leisure. Nordhaus and Tobin (1972) added the value of leisure time to the value of consumption in each year evaluated at the real wage rate of a base year. They also made allowance for the possibilities *either* that 'productivity' of workers in making use of leisure time remained unchanged *or* that it rose with labour productivity in general.

No one has good evidence on this last point although participants in the standard of living debate have implicitly discussed it. For example, pessimists tend to argue that for urban workers opportunities to enjoy leisure were much less than they had been in the countryside (Hammond, 1930) whilst optimists emphasize rising literacy (Hartwell, 1961).

Actually, *unless* it is assumed that productivity of workers in using leisure time rose, the Nordhaus and Tobin method of valuing changes in leisure time would surely lead to a very low implied rate of growth of overall consumption. For example, assuming zero productivity change, and valuing leisure time at the real wage rate of 1851, gives 0.10 per cent per year growth in living standards adjusted for leisure for 1760–1851. Again, the point emerges that items neglected by national-income accounting may dominate conventional consumption in this period and that there is no consensus on how to deal with these problems.

6

Structural Change

I INTRODUCTION

In Chapter 3 we saw that Britain had a different pattern of structural change from that experienced on average in Europe during the nineteenth century. Most striking of all was the very rapid decline in agriculture's share of the labour force and the achievement early on of parity between output per worker in agriculture and the rest of the economy. Thus, by 1840, when Britain's income level was $567, the percentage of the male labour force in agriculture was 28.6 per cent compared with a 'European Norm' for that income level of 54.9 per cent, and the share of income in the primary sector was 24.9 per cent compared with a 'European Norm' at that income level of 37.2 per cent. In Chapter 4 it was estimated that total factor-productivity growth was generally quite rapid in agriculture, and at times exceeded that of manufacturing and the rest of the economy—most notably during 1801–31 when agricultural productivity growth was estimated at 0.9 per cent per year compared with 0.3 per cent for manufacturing. In this chapter it is time to relate the results of Chapters 3 and 4 a little more closely, in particular to examine the conditions which generated this unusual transformation of the economy.

The recent historiography of the British industrial revolution makes a closer look at the changing relative share of agriculture in the economy particularly desirable. For a long time the literature stressed that in many ways agricultural techniques remained labour intensive, and that total labour inputs in agriculture rose over time and appeared to deny that agriculture released labour to industry during the industrial revolution. Jones sums up one of the best known surveys as follows: 'It would be tendentious to praise agriculture because its inability to release enough labour prompted inventiveness, but it must be concluded that it was not usually an immediate source of labour for industry' (1974, p. 102), and Timmer concluded that 'The English agricultural revolution increased land, not labour productivity' (1969, p. 392). Moreover, textbooks tended to assert that until the 1840s the contribution of imports to the feeding of Britain's population was very small (Chambers and Mingay, 1966, p. 208; Mathias, 1983, p. 71). Set next to the very low shares of agriculture in the labour force and income in 1840 this account of agriculture's role in Britain's industrialization seems rather paradoxical.

This chapter argues that the preceding textbook accounts are in need of some revision. In Section II a simple 'ex-post accounting' model is used to show that in the economist's usual sense of the term agriculture did release labour to industry and that the productivity growth discussed in Chapter 4 was underwriting this process. In Section III the transfer of resources between sectors is considered further, and it is suggested that a net flow of savings from agriculture was made available to industry and transport. This picture is essentially in line with conventional accounts. In Section IV, however, I revert to revisionism and show that on the basis of recent research it is clear that imported food was of great importance in the first half of the nineteenth century; Britain became a more substantial exporter of manufactures (notably textiles), and the net barter terms of trade declined substantially. It is argued that a pattern of increasing specialization in exportables helps to explain Britain's structural transformation even prior to the abolition of the Corn Laws. The sources and nature of that specialization are further examined in Chapter 7.

II RELEASE OF LABOUR BY AGRICULTURE

We can start with a model which is very simple but can offer useful insights into the experience of 1700–60, and serve also as an interesting benchmark case. Consider an economy whose food is entirely domestically produced and in which the relative price of food is actually sustained at the same level, although in principle free to vary. Suppose agriculture is characterized by diminishing returns to labour, and suppose also that food demand grows at the same rate as population. If the relative share of the labour force in agriculture is to fall, that is, industrialization is to occur, it is necessary that the growth of the labour force in agriculture be less than the growth rate of the labour force overall, which we will assume is equal to the growth rate of the population. This appears to pose problems, for whilst the demand for food grows with population the food supply grows less rapidly even if all the extra labour is used in agriculture (which would, of course, amount to a de-industrialization). Clearly to home feed all the extra population and also to industrialize by having the agricultural labour force growing less rapidly than the labour force as a whole it would be necessary to have output per worker in agriculture rising. This in fact is the crucial condition for what Dixit (1973) has termed the 'viability' of the economy. Obviously, if rising per capita incomes were to generate food-demand growth faster than population growth the preceding arguments hold *a fortiori*. Since it is supposed that there are diminishing returns to labour in agriculture, industrialization requires some other force, such as technical progress or capital accumulation, to raise output per worker in agriculture by more than offsetting the diminishing returns.

This argument can be formalized as follows. The production function in agriculture is assumed to be

$$Q_t = Ae^{\mu t}L_t^{\lambda} \quad (0 < \alpha < 1) \tag{6.1}$$

where μ represents the rate at which capital accumulation and technical progress are augmenting output. The rate of growth of food supply is therefore taken to be

$$\dot{Q}/Q = \mu + \alpha.\dot{L}/L \tag{6.2}$$

If prices remain constant, then the growth of demand for food must be equal to the rate of growth of supply:

$$\mu + \alpha\ \dot{L}/L = \xi\ (\dot{Y}/Y - n) + n \tag{6.3}$$

where ξ is the income elasticity of demand of food and n is the rate of population growth. This can be rearranged to give the required rate of growth of the agricultural labour force to meet the growth in demand for food as

$$\frac{\dot{L}}{L} = \frac{\xi\ (\dot{Y}/Y - n) + n - \mu}{\alpha} \tag{6.4}$$

For industrialization $\dot{L}/L$ has to be less than n. If this condition is met, then the economy can be said to be viable in the sense that the food requirements of the extra population can be met by a rate of increase of the labour force in agriculture less than the rate of increase of the labour force and population as a whole. Then some labour is released for use in industry and the food requirements of the population can be met by a lower share of the labour force. This release of labour requires

$$\frac{\xi(\dot{Y}/Y - n) + n - \mu}{\alpha} < n \tag{6.5}$$

which can be rearranged to give

$$\mu - (1 - \alpha)n - \xi\,(\dot{Y}/Y - n) > 0 \tag{6.6}$$

Thus even if there were no increase in income per head, with population growth for viability we need $\mu > 0$ because $\alpha < 1$. Given α, the faster is population growth the higher needs to be μ. That is, industrialization would require that capital accumulation and technical progress in agriculture more than outweigh diminishing returns to agricultural labour so that output per worker in agriculture rises. Moreover, the necessary μ for industrialization would tend to be raised for faster growth of income per head and/or a higher income elasticity of demand for food.

Note also that equation (6.4) shows that to get an absolute decline in the agricultural labour force we need $\mu > \xi\,(\dot{Y}/Y - n) + n$, which is a more stringent condition than meeting inequality (6.6). It is therefore quite possible to observe an industrialization of the labour force in the sense that (6.6) is met and the economy is viable, whilst the agricultural labour force is growing in absolute size. Properly conceived, the release of labour from agriculture to industry is concerned with a decline in the proportion of the labour force in agriculture, not a decline in absolute numbers. The important thing is the ability of the economy to meet the extra food requirements occasioned by the growth of population and incomes whilst allowing the share of industry in the labour force to rise.

It is easy enough to take a look ex-post at the economy during 1700–60 in terms of this model. As we saw in Chapter 2, prices in 1760 were the same as in 1700 and trade in foodstuffs was relatively unimportant. The data of Chapter 2 can easily be recast in terms of our simple model. The additional piece of information required is the value of α; Ippolito (1975, p. 308) gives an estimate of 0.36 per cent and in Chapter 4 we saw that the share of labour in agricultural output was 40 per cent—a value of 0.4 seems plausible for α. From Chapter 2 for 1700–60 we have an estimate for $\dot{Q}/Q$ of 0.60 and for $\dot{L}/L$ of -0.05. Using equation (6.2) this gives an estimate for μ in 1700–60 of 0.62.

Recalling the evidence of Chapter 2 it was also estimated that the share of the labour force in agriculture fell between 1688 and 1759 from 55.6 per cent to 48.0 per cent. Thus the economy was viable and labour was released by agriculture during this period, prior to the classic industrial revolution; output per worker in agriculture rose substantially even though absolute numbers in agriculture fell only slightly. We have already seen that the high positive value for μ was based especially on increasing yields per acre as better crop rotations became more widely adopted.

Before leaving this elementary example it is worth exploring two counterfactuals, which reveal the value of the simple model. First, we can consider the impact of the value of ξ, the income elasticity of demand for food. Chapter 2 already gave us an estimate of 0.7 for ξ; this appears to be a very normal value for ξ (FAO, 1962) and is by no means the unusual Japanese case where ξ was perhaps only 0.2 (Kelley and Williamson, 1974, p. 160). As Table 6.1 indicates, the μ required for viability would have been very substantially reduced had ξ been only 0.2, and it would be inappropriate to explain the unusual structural change in Britain on account of a low income elasticity of demand for food.

Secondly, we can consider the impact of a different rate of population growth. This is of particular historiographic interest. The approach of many economic historians was greatly influenced by the work of Chambers (1953). Chambers argued that the expansion of the industrial labour force came about because 'the movement of population had taken an upward turn in village and town alike and provided an entirely new supply of human material' (1953, p. 338). This argument soon reached the status of orthodoxy and an even bolder form as, for example, is shown by the following quotation; 'the agricultural revolution associated with the enclosures increased the demand for farm labour . . . the rapid growth of population created a surplus of labour in the countryside much of which found its way into the new urban centres' (Landes, 1969, pp. 115–16).

Table 6.1 reveals that this argument is seriously misleading, at least in the simple form argued above. Part (*c*) of the table shows that in order to retain constant prices if *n* were zero it would only

Table 6.1. Release of Labour from Agriculture: Some Permutations on the 1700–60 Experience

---

(*a*) *Viability* in the simple model requires

$$\mu - (1 - \alpha)n - \xi(Y/Y - n) > 0$$

(*b*) *Estimated Values for 1700–60.*

$n : 0.38$	$\dot{Q}/Q : 0.60$	$\xi : 0.7$
$\dot{Y}/Y - n : 0.31$	$L/L : -0.05$	$\mu : 0.62$

(*c*) *Required* μ *for Viability.*

(i) $\xi = 0.7 : 0.45$
(ii) $\xi = 0.2 : 0.29$
(iii) $n = 0.0 : 0.22$
(iv) $n = 1.5 : 1.12$

---

be necessary to have $\mu = 0.22$. Conversely, at a population growth rate of 1.5 per cent the required μ rises sharply to 1.12. Thus, higher population growth rates place greater strains on the economy's ability to industrialize rather than of themselves directly promoting industrialization in the closed economy case. A more general statement of this argument is given in Crafts (1985c).

In fact, industrialization did involve greater strain on agriculture as both population growth and growth of incomes per head eventually rose much higher than in 1700–60. Thus, whilst by 1820–40 μ had risen to 1.27, the required μ for viability in the simple model had risen to 1.54 and the μ required for absolute declines in the agricultural labour force had risen to 2.07 (Crafts, 1985c). This high value of μ at 1.27 would, however, have been consistent with viability at a rate of growth of income per head of 0.65 per cent per year and thus the economy was capable of a release of labour. Probably only in the period of slow agricultural growth in the late eighteenth century (see Chapter 2) was this not so.

After 1760, however, the behaviour of the economy no longer approximates our simple model. Prices for agricultural products rose considerably and, as Section IV shows, in the early nineteenth century imports had become very important in Britain's food supply. Yet it is important to bear in mind that the economy generally had the capability for a release of labour from agriculture. In fact, the literature has in *effect* always accepted this (and even possibly overstated it) without conceptualizing the notion of labour release clearly. For example, Jones and Woolf conclude that 'labour was probably not released from agriculture during the first wave of change—mixed farming had heavy labour needs and the absolute number of farmhands actually grew—but the nation's food supply could be secured by an ever smaller proportion of the national workforce' (1969, p. 15). This last statement admirably sums up 1700–60, and our simple model has revealed that this *was* a release of labour.

Finally, it is important to be clear that the low share of agriculture in income and the labour force by 1840 was based on levels of efficiency which were very high by European standards. Chapter 4 showed that total factor productivity growth in agriculture was rapid after 1800 and impressive prior to 1760. Table 6.2 reports Bairoch's calculations of levels of labour productivity in 1840; Britain has a high productivity level and, for example, emerges as superior to Germany which was an important supplier of grain to Britain in the mid-nineteenth century.

Table 6.2. Physical Productivity of Labour in Agriculture in 1840 (million direct calories per male agricultural worker)

Britain	17.5
France	11.5
Belgium	10.0
Switzerland	8.0
Germany	7.5
Sweden	7.5
Russia	7.0
Italy	4.0

Source: Bairoch (1965, p. 1096).

This section has shown that the economy generally had the ability to cope domestically with increased demand for food and release labour from agriculture without experiencing food-price rises, whilst achieving a respectable, but not high, rate of growth of per capita income. This has not been fully recognized in most of the literature, and as a result the low share of agriculture in output and the labour force which was characteristic of the early nineteenth century has appeared somewhat puzzling. In fact it was based domestically on high agricultural-labour productivity and an income elasticity of demand for food of less than one. In addition, as Section IV discusses, the economy moved to a pattern of international specialization based on importing food in the early nineteenth century.

III SECTORAL SAVINGS FLOWS AND INDUSTRIAL CAPITAL ACCUMULATION

Structural change was based, of course, not just on a reallocation of labour but also on a change in the composition of investment. Thanks to the research of Feinstein (1978) this redirection is now well quantified. Feinstein's results are summarized in Table 6.3.

Table 6.3. Gross Domestic Fixed Capital Formation by Sector
(£m, annual averages, current prices)

	1761–70	*1791–1800*	*1821–30*	*1851–60*
Agriculture	1.20	3.41	4.00	6.90
Industry and Commerce	0.77	2.75	11.50	20.67
Transport	0.84	2.45	4.45	18.12
Residential and Social	0.87	2.80	11.38	12.30
Total	3.68	11.41	31.38	57.99

Source: Feinstein (1978, Table 7).

In Table 6.3 we see agriculture accounting for 32.6 per cent of gross domestic fixed capital formation in 1761–70, but only 12.8 per cent in 1821–30. The major decline in agriculture's share of capital formation occurred between 1791–1800 and 1821–30. During 1800–60 output growth in agriculture was considerably higher than capital-stock growth, which was around 0.7 per cent per year (Feinstein, 1978, p. 68), although in turn this nevertheless implied a steadily rising capital-to-labour ratio in agriculture. Finally, it should be noted that Feinstein estimated that the proportion of rents going to investment in agriculture rose from 6 per cent in 1761–70 to 16 per cent in 1801–10, after which it declined very slightly (1978, p. 49).

Thus, in the early nineteenth century agriculture's potential to release labour was based on total factor productivity growth accelerating (Chapter 4), but *also* involved a growing capital stock based on a higher proportion of rents being reinvested in the sector. In fact, it was not until 1811–20 that fixed capital formation in industry and commerce exceeded that in agriculture.

These circumstances have led to considerable uncertainty as to how far agriculture transferred funds for investment to the rest of the economy. Jones nicely captures the general tendency of the literature:

> For all the serious rigidities of the capital market in the countryside, all the feedback of investment by industrialists keen to acquire the social colouring of landed society, and all the expense of enclosure, the balance of probabilities is that agriculture did make a net contribution to the formation of industrial capital and did release entrepreneurs who played a significant role in industrialization. (1974, p. 109).

To some extent the uncertainty probably reflects a confusion between flows of investment and changes of ownership of existing assets. A purchase of land by a successful industrialist is an example of the latter; the change of ownership of the assets may subsequently lead to altered sectoral or total savings propensities, but is not of itself a flow of investment spending creating final capital goods.

The questions of interest are to what extent the savings propensities from different income flows varied in a given year, and also over time, and whether there was an excess of savings in agriculture over investment spending on new capital goods for use in agriculture. This can be put in terms of elementary macroeconomic income accounting. In a year with a balanced government budget and zero balance of payments on current account we have

$$S \equiv I \tag{6.7}$$

$$S_a + S_{na} \equiv I_a + I_{na} \tag{6.8}$$

$$S_a - I_a \equiv I_{na} - S_{na} \tag{6.9}$$

$$saY_a - I_a \equiv I_{na} - s_{na}Y_{na} \tag{6.10}$$

$$saY_a - I_a + s_{na}Y_{na} \equiv I_{na} \tag{6.11}$$

where the subscripts *a* and *na* represent agriculture and non-agriculture respectively. Equation (6.11) indicates that the savings flows to match investment flows outside of agriculture depend on savings within the non-agricultural sector and the excess of savings in agriculture over investment in agriculture. In looking at increases in non-agricultural investment over time, in principle we would like to know to what extent s_a and s_{na} altered over time, to what extent s_aY_a grew faster than I_a, and if s_a and s_{na} were not equal whether the tendency for Y_{na} to rise relative to Y_a over time added to total savings.

Unfortunately there is insufficient information to measure s_a and s_{na}, the propensities to save in the two sectors. It was established in Chapters 4 and 5 that the economy's overall propensity to save rose during the industrial revolution, and that the savings came from rents and profits rather than wages. The macroeconomic information is also sufficient to establish that the rise in the overall savings propensity involved changes in one or both of the sectoral savings propensities and not just a change in sectoral income shares.[1]

Beyond this we are in the realms of conjecture. It seems quite probable that, at least by the 1820s, saving in agriculture exceeded investment in the sector. For example, if we assume that agriculture saved enough simply to finance its own investment and that saving out of wages was zero, then savings would have been 29.4 per cent of non-agricultural non-labour income[2], a proportion which seems somewhat high. We also have good reason to believe that agricultural savings, on occasion at least, exceeded agricultural investment. Whilst Chambers and Mingay argue that parliamentary enclosure, the major undertaking in agriculture, was almost entirely financed within agriculture (1966, p. 83), Ward (1974, p. 74) found that landed gentlemen and farmers contributed 18.9 per cent of the share capital of the canal promotions that he studied.

Writers studying investment in this period have tended to assume that reinvestment of profits was the major single source of finance for industry, but that the reinvestment fraction was not enormously high. Thus O'Brien, in studying the role of profits from the 'periphery' in the industrial revolution, considers that a 30 per cent reinvestment fraction would only be true of the 'exceptionally frugal' (1982b, p. 7). Nevertheless, the average propensity to save out of non-labour income was about 25 per cent, and if we also take O'Brien's figure of a 30 per cent upper bound for savings out of rents and profits in agriculture, then the

[1] Using the data on factor shares assembled in Chapter 4 it would seem that non-labour income in agriculture was about 22 per cent of national income in 1760, and 13 per cent in 1840, whilst profits in the rest of the economy were about 33 per cent and 42 per cent respectively. Overall savings had risen from 11.6 per cent of non-labour income to 22.4 per cent. Assuming constant s_a and s_{na} and solving the simultaneous equations system gives the nonsense result that s_a = *minus* 0.3.

[2] From Feinstein (1978, Table 7; and Deane and Cole, 1962, p. 166) we have an overall savings propensity of 13.8 per cent. Profits in non-agriculture represented about 39 per cent of national income, and agricultural savings by assumption were 2.2 per cent of national income; hence the calculation in the text.

propensity to save out of non-agricultural non-labour income would not have been less than 23 per cent. Equally, a 30 per cent savings rate from agricultural rent and profits would have generated in 1821–30 a savings flow net of agricultural investment of £7.5m., which is only just over one fifth of the home and foreign investment taking place outside agriculture.

It is not possible, with present data availability at least, to suggest a narrower range of possibilities than this. In sum, it seems likely that there was a positive net flow of investment out of agriculture but that this was not more than a small fraction, at most a fifth, of the total flow of non-agricultural investment.

Before leaving the topic of savings flows available to the economy to finance the expansion of the non-agricultural capital stock, it is also opportune to briefly consider a further flow, from trade to industry. There is a tradition in British economic history which, although much challenged recently, argues that Britain's dealings with countries in the Empire or peripheral to Europe generated a stream of profits which was of great importance to industrial expansion. Williams provided the most famous statement: 'there was hardly a town in England which was not in some way connected with the triangular or direct colonial trade. The profits obtained provided one of the main streams of that accumulation of capital in England which financed the Industrial Revolution' (1944, p. 52).

O'Brien has made calculations which indicate that this is an exaggeration. Using a 30 per cent reinvestment rate he calculates an upper bound: 'commerce with the periphery generated a flow of funds sufficient, or potentially available to finance about 15 per cent of gross investment expenditures undertaken during the Industrial Revolution' (1982b, p. 7).

Together with our earlier discussion of agricultural savings and investment flows, this seems to add up to a picture of at least two-thirds of industrial investment being financed out of savings from non-labour income generated on domestic non-agricultural activity other than trade with the periphery.

IV INTERNATIONAL TRADE AND THE STRUCTURE OF HOME PRODUCTION

For countries engaging in international trade the structure of domestic production and consumption differs; this was, of course, true for Britain even before the move to Free Trade in the 1840s. The availability of trade statistics in current prices resulting from the work of Davis (1979) when placed together with estimates of home production levels illustrates this point much better than was possible hitherto. The figures in Table 6.4. illustrate the nature of trade in the first half of the nineteenth century; there was a substantial excess of production over consumption in manufactures, which was exported, whilst in foodstuffs there was a substantial excess of consumption over production, which was imported.

First of all, Table 6.4 indicates that by the beginning of the nineteenth century about one sixth of all food consumption was of net imports. In 1841 the proportion was 22 per cent, rather similar to what it had been in 1821. The abolition of the Corn Laws, of course, raised the share of net imports in food consumption; in 1851 Table 6.4 indicates that the share had risen to 28.5 per cent. The emphasis of the textbooks on home-feeding during the industrial revolution is somewhat overstated; as Thomas argues, the reasons for this appear to be failure to note the important role of Ireland as a British supplier, which is obscured by the trade statistics after 1825, and an overemphasis on bread and meat in discussions of food and too little note of tea, sugar, coffee, etc. (1984).

The importance of food imports should not connote agricultural inefficiency in an absolute sense, or an inability of agriculture to release labour; Section II made it clear that these were not correct diagnoses. On the contrary, the economy was exploiting comparative advantage both before and after the move to Free Trade, more fully so after the move to Free Trade. Comparative advantage is, of course, based on *relative* efficiencies, and trade allowed the economy to achieve higher levels of welfare than would have been the case under autarchy.

The second point that can be derived from thinking about Table 6.4 is that the specialization in production of manufactures which it reveals has implications for the overall structure of economic activity as compared with an economy less committed to international trade. In order to specify precisely what participation in trade implied for structural change would require a complex general equilibrium model much too demanding for the available

Table 6.4. Domestic Production and Consumption, 1801, 1841, and 1851 (£m)

	Manufacturing, Mining, and Building	*Agriculture*
1801		
Home Production	54.3	75.5
Net Exports(−)/Imports(+)	−22.6	+14.8
Home Consumption	31.7	90.3
1841		
Home Production	155.5	99.9
Net Exports(−)/Imports(+)	−32.5	+28.9
Home Consumption	123.0	128.0
1851		
Home Production	179.5	106.5
Net Exports(−)/Imports(+)	−51.1	+49.6
Home Consumption	128.4	156.1

Source: Derived from Deane and Cole (1962, p. 166) and Davis (1979, pp. 96, 100, 101, 104, 108, 109, 114, 115, 122–5). Agricultural imports were adjusted to include Ireland following the method of Thomas (1985).

data, but a procedure used by McCloskey (1981b, ch. 7) can be adapted to give a crude estimate.

France in 1840 had a ratio of exports to GNP of about 6 per cent,[3] compared with a ratio of about 13 per cent in Britain, proportions which were fairly typical of the thirty years after the Napoleonic Wars. Following McCloskey's procedure, suppose that geography and trade policy had conspired to reduce the trade ratio in Britain to a similarly low fraction of national output as in France, that net exports and imports were reduced pro rata, and that domestic production adjusted so that consumption remained unchanged. (In practice, we could expect consumption to take some part of the adjustment as relative prices would change.) Table 6.5 displays the results of this counterfactual calculation.

Table 6.5 suggests that reduction of trade to the French level might have quite an interesting effect on domestic economic activity. Reference to Table 3.6 shows that the counterfactual change of 3.9 percentage points in the share of industry in total output would drive the share of industry towards the 'European Norm' for this income level, although agriculture's share would still be about 9 percentage points below the 'European Norm'.

[3] Calculated from Kuznets (1967, p. 99) and Caron (1979, p. 11).

Table 6.5. Counterfactual Calculations of Primary Sector Shares in Output and the Labour Force in 1841

	(1)	(2)	(3)	(4)
Output Share	24.9	28.3	27.2	37.2
Labour Force Share	25.0	28.4	41.2	53.7

Notes.
Column (1): actual data from Table 3.6.
Column (2): counterfactual described in the text in which trade is reduced to French proportions.
Column (3): a more complicated counterfactual in which not only is trade reduced to French proportions but *also* labour productivity in agriculture is set at French levels, as in Table 6.2, and labour is reallocated to prevent any fall in agricultural output.
Column (4): the European Norm from Table 3.6.

This counterfactual can be enlarged upon by considering the further ramifications for labour use. If labour-to-output ratios were unchanged, then the rearrangement of production could be expected to raise the share of the labour force in the primary sector by about 3.4 percentage points, only a small fraction of the gap between the actual and European norm.

Obviously this is a crude calculation, and the actual implementation of a move to an economy less involved in trade might have involved many more complicated interactions in the economy. Nevertheless, from Table 6.5 we get an approximation of the impact of specialized production for trade on the structure of the British economy. There seems to be a case to suggest that specialization in production for export, to a greater degree than in France, could account in large part for the unusually high share of industry in production, but not for the low share of agriculture; Britain had a large services sector. Still less does the relatively large excess of consumption over production in British agriculture, as compared with French agriculture, account for the very low share of the labour force in British agriculture in 1841; this seems to owe much more to relatively high output per worker in British agriculture, (release of labour) as Table 6.5 also shows.

The above counterfactual seems to be quite an interesting one in the sense that whilst it would be quite unrealistic to consider Britain as an autarchic economy in this period, tariff policy was variable and did influence the extent of trade, most famously with the move to free trade in the 1840s. Much more speculative counterfactuals have been introduced into the industrial revolution literature and a vigorous debate has taken place, particularly with regard to the second half of the eighteenth century. Thus, for example, Hobsbawm's classic statement concerning the origins of the industrial revolution is that 'Home demand increased but foreign demand multiplied . . . If a spark was needed, this is where it came from.' (1968, p. 32). Barrett-Brown, supporting this position, laid great stress on trade as an 'engine of growth', and drew attention to the rise in the export/GNP ratio for the economy during the industrial revolution (1974, pp. 103–11). By contrast Hartwell thought that 'The steadily expanding level of aggregate domestic demand was more important for growth than the more erratic growth of foreign trade' (1971, p. 152) at least before 1780, a position shared by John (1965) and Eversley (1967), who both stress the increasing purchasing power of the home market and, during much of the eighteenth century, agriculture. Deane and Cole after a lengthy discussion concluded that 'it is difficult to see the expansion of the British export trade as a largely exogenous factor which quickened the pace of industrial growth . . . it seems that the explanation of the higher average rate of growth in the second half of the century should be sought at home rather than abroad' (1962, p. 85).

The debate is about the relative importance of export markets in accounting for British industrialization, both in terms of the growth of industrial output and the move towards a higher share of industry in total output and income, the chief concern of this chapter. It seems to raise three issues which are worth discussing in addition to the analysis of Tables 6.4 and 6.5.

First, and most simply, the literature requires an answer to the question of the ex-post importance of different flows of expenditure on total output and on industrial output. Second, we are asked to consider whether changes in external conditions played a major part in promoting overall growth. Third, and most difficult, the question is raised, under what combination(s) of circumstances at home and abroad would structural change have been less rapid at any given rate of growth?

The first question has recently been reviewed extensively by Cole (1981), by Crouzet (1980), and by O'Brien (1982a). O'Brien concluded that the growth of agricultural incomes accounted in fact for 29–44 per cent of the expanded sales of industry during the eighteenth century, and thus agriculture was not the primary source of increases in demand for industrial output (1982a, pp. 24–8). Cole concluded that as far as industry was concerned nearly 40 per cent of additional output was sold as exports (1981, p. 40). Crouzet came to the most dramatic conclusion, reviewing national output as a whole rather than just industry. He found that 'during most of the century up to the 1780s, the ratio of domestic exports to national income was quite modest and remained under 10 per cent' (1980, p. 77), but that 'During the decisive stage of the Industrial Revolution in the twenty years which followed the peace of 1783, the incremental ratio of exports to national income seems to have been as high as 40 per cent' (1980, p. 82).

The new estimates of growth indicate that Crouzet's estimates are rather misleading, particularly for 1783–1801. Crouzet appears

Table 6.6. Exports as a Proportion of National Output and Its Growth (%)

Exports/National Output		*Increase in Exports as a Proportion of Increase in National Output*	
1700	8.4	1700–60	30.4
1760	14.6	1760–80	5.1
1780	9.4	1780–1801	21.0
1801	15.7	1801–31	11.3
1831	14.3	1831–51	29.4
1851	19.6		

Sources: National Output is based on Deane and Cole (1962, p. 166) for 1801–51 and on Chapter 2 for the eighteenth century. The figure for 1700 is tentative. Exports are based on Deane and Cole (1962, p. 321) for 1700–1801, adjusted for price changes using Davis (1979, p. 86) for 1780 and 1801; for 1831 and 1851 figures are from Davis (1979, p. 86) using the 1854–6 figures for 1851 and adjusting the 1834–6 figures for the omission of Ireland pro-rata with 1824–6. All calculations in this table are based on current prices.

Note: These figures are slightly corrected from those given in Crafts (1983a, p. 197).

to have used too high a value of national income in current prices and has a biased estimate as a result;[4] his export figures are taken from the same sources as underlie Table 6.6, and are perfectly acceptable. Table 6.6 displays evidence on exports relative to national output in current prices. In the long run the picture is not dissimilar to Crouzet's, but the short-run picture in the late eighteenth century is very different. Growth of exports relative to growth of national output in that period is, of course, distorted by war, but in any event does not appear to have been dramatically large. It should also be remembered that the export figures are for gross output not value-added, and that Table 6.6 somewhat inflates the importance of exports as an ex-post source of incomes. Nevertheless, exports as a share of total output clearly did rise during the industrial revolution over the long run, although on occasions the expansion path was an erratic one.

Table 6.7 breaks new ground in an attempt to look at the uses of industrial output alone, perhaps an exercise pertaining rather more to Hobsbawm's claim. This table requires rather more detailed discussion that the previous one, partly because its results must be regarded with caution—at least for columns (4)–(6).

[4] Crouzet used a figure of £160m. for national income in 1783, compared with £232m. in 1801. This would not tally either with Deane and Cole's output index and what is known about price changes or with the new estimates. The new estimates give a national income of £105m. in 1780.

Table 6.7. Purchases of Gross Industrial Output in Current Prices (£m.)

	(1) Gross Ind Y[a]	(2) Exports[b]	(3) Investment[c]	(4) Workers[d] Consumption	(5) 'Capitalists[e] Consumption'	(6) Agriculturalists[f] Uses
1700	15.6	3.8	1.9	3.0	6.9	4.3
1760	23.6	8.3	3.4	4.8	7.1	4.5
1780	39.9	8.7	6.6	6.5	18.1	9.8
1801	82.5	28.4	14.6	13.0	26.5	23.4
1831	178.0	38.9	35.6	18.7	84.8	29.0
1851	272.8	67.3	52.9	30.9	121.7	39.6

[a] Assumed to be equal to 1.52 times value-added; based on the estimates of Chapter 2 for 1700, 1760, and 1780, and Deane and Cole (1962, p. 166), for 1801, 1831, and 1851.

[b] Taken from Davis (1979, pp. 94–101) for 1780–1851 using averages of adjacent years, e.g. 1851 is the average of 1844–6 and 1854–6 figures. For 1780 adjusted downwards from 1784–6 pro rata with Deane and Cole's figures for all exports in constant prices (1962, p. 321). For 1700 and 1760 exports are from Davis (1969, p. 120), adjusted by the ratio of British to England and Wales exports in 1775–84 from Deane and Cole (1962, p. 48).

[c] Taken from Feinstein (1978, p. 41, 69) using averages of adjacent decades, except for 1760 which is assumed equal to 1761–70; for inventory investment proportional to industry's weight as a fraction of industry, commerce, and agriculture. Industry is assumed to supply all investment goods except in agriculture, where it is assumed to supply 50 per cent of investment goods. For 1700 investment is taken to be 4.0 per cent of national expenditure all supplied by industry; national expenditure is based on a figure of £48.0m. Crafts (1983a, p. 189).

[d] Workers consumption is based on an expenditure of 12.5 per cent of wages for 1700, 1760, and 1780. Wages were assumed to be 50 per cent of national income. The share of 12.5 per cent allows for a mark of up to 100 per cent in distribution, transport costs, etc. and would imply 25 per cent of workers budgets went on industrial output cf Ch. 5. For 1801, 1831, and 1851 wages are from Deane and Cole (1962, p. 152).

[e] Capitalists consumption is obtained as a residual and is equal to column (1) – (col. (2) + col. (3) + col. (4)). Obviously it includes 'landowners' and 'rentiers' consumption also, and government current consumption of industrial goods.

[f] This is obtained using Feinstein's figures for fixed capital formation in agriculture divided by two, *plus* the assumption that workers spent 7.5 per cent of wages (= 40 per cent of agricultural income) on industrial goods, *plus* the assumption that the fraction of the 60 per cent of income in agriculture going to profits and rents was equal to the fraction of 'capitalists consumption' in column (5) in non-labour income in the economy as a whole.

Not surprisingly, since total national output includes many items that we know did not enter international trade, exports are a higher proportion of industrial output (even though the calculations are made in gross terms) than of national output. The proportion in Table 6.7 ranges from 22 per cent in 1780 and 1831 to 35 per cent in 1760. In the long run, however, the proportion of exports to gross industrial output was little different in 1700 and 1851. Cole's finding is confirmed in that Table 6.7 gives an estimate that 36.8 per cent of increased industrial output was in the form of exports, taking the eighteenth century as a whole. This proportion falls to just above 20 per cent in 1801–51, however, as prices of British exports fell substantially although volume increased steadily (this point is elaborated in detail in Chapter 7). In this period home investment accounts in current prices for an increase in sales of gross industrial output fairly similar to that coming from exports.

It is noteworthy that pessimists in the standard of living debate have always tended to stress the importance of export markets in British industrialization (cf. Hobsbawm above). Table 6.7 perhaps helps to explain why this is so.[5] Column (4) gives estimates of increases in workers' consumption of gross industrial output; as the notes to the table indicate, workers' consumption is derived from assumed percentages of labour income spent in industrial output, a proportion broadly consistent with what we know from budget studies (see Chapter 5). The calculation may be challenged in detail, but no reasonable change in assumptions would make workers' consumption anything other than a very small fraction of the market for industrial output at any time during the industrial revolution. Equally clearly, home consumption other than that of workers was generally the most important single outlet for industrial output.

In terms of the historiography column (6) is also very interesting. From the discussion in Section III of this chapter it seems likely that the assumptions underlying this column are somewhere near the mark, although accuracy in complete detail is not possible, of course. The results are consistent with O'Brien's estimates; it emerges that for the eighteenth century the point estimate from

[5] Table 6.7 is in current prices and does not indicate changes in real wages. These were discussed in Chapter 5. Table 6.7 is another way to throw light on the standard of living debate, but does not supplant the discussion of real consumption in Chapter 5.

Table 6.7 is that 28.6 per cent of incremental industrial output was sold to agriculturalists for investment *and* consumption purposes. This is less than the estimate of the proportion going to exports. As O'Brien implies it appears that writers such as John, whilst correct in describing steady growth in agriculture, may have exaggerated somewhat that sector's importance as a purchaser of industrial output. Not surprisingly from 1801 onward agriculturalists represent a rather small percentage of extra industrial output sales, probably less than 10 per cent of the total from 1801–51.

Whilst tables 6.6 and 6.7 give the best available answers on the first of our three issues, it is important to recognize that such evidence does *not* constitute an adequate response to the other two questions posed. This becomes clear immediately serious consideration is given to the role external conditions may have had in promoting overall growth.

The arguments of the protagonists in the debate touched on above concerned sources of increased demand for the economy. Neo-classical models of the growth process would in any event find this an uninteresting argument because in such approaches growth is determined on the supply side of the economy and demand plays a passive role (Mokyr, 1977). If, perhaps for shorter periods of analysis, demand is given an independent (Keynesian) role, then the key analytical question to be considered is which increases in demand are taken to be *exogenous*, that is, independent, stimuli to growth. Such a question cannot be answered at the simple level of accounting we have been considering. It should be noted, however, that personal consumption expenditure would *not* be exogenous in a Keynesian demand model[6] and that most likely for short-run analyses the relevant candidates to be considered as exogenous demand stimuli would be exports, investment, and government spending. As comparison of Tables 3.6 and 6.6 shows, in general exports *were* the largest of these three categories of expenditure, at least in peacetime. Recent work by Hatton *et al.* (1983) has strengthened the claims of exports to be considered as exogenous in short-run models of the eighteenth-century economy, although, as Crafts (1981) shows, the relative importance of exports as a source of exogenous increases in demand varies according to the period, being particularly weak in 1760–80.

[6] Consumption would be taken to be a function of income and to be endogenous.

In the longer run a further possibly important impact of external conditions on overall growth also has to be considered, namely the contribution of the terms of international trade (basically the price of British exports compared with the price of British imports) to changes in real income. The ability of the economy to avoid 'immiserizing growth', in which domestic increases in productive potential dramatically raise the relative price of imports/exports, involves external conditions, is important, and has been little discussed. Tables 6.6 and 6.7 merely subsume some of the effects of changes in the terms of trade. This issue requires a full discussion which can be found in Chapter 7.

In answering the first question posed, 'who bought industrial output?' Table 6.7 compared exports and agricultural demands, as has been commonplace among earlier writers. Agricultural demands were based on agricultural incomes and would be higher the more that sector grew, and thus a long tradition in the literature has been to contend that faster agricultural growth stimulated industrialization by providing a bigger market for industrial output. By far the most careful examination of this argument was made by O'Brien, who concluded that for the eighteenth century exports should be given greater prominence than increases in agricultural productivity (1982a, p. 26). It is important to realize, however, that this does *not* imply that had agricultural growth been greater industrialization would have been faster.

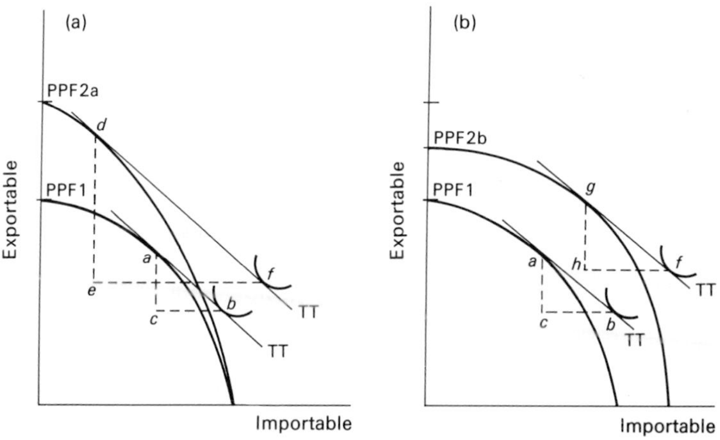

Figure 6.1. Production Possibility Frontiers and the Composition of Output

In fact, such counterfactual propositions need to be considered in the context of Britain's participation in international trade where, as we have seen, manufactures represented exportables and agricultural goods represented importables. This perspective has not generally been used in the literature, although, as Section II above noted, at least from 1760 on this appears to be an important omission.

An aid to thinking about the pace of industrialization in a trading economy is provided by Figure 6.1. In both parts of this figure diagrams are drawn of production possibility frontiers for a two-good model of the economy. The figures are drawn for a small country for whom the terms of trade are constant and represented by the slope of the (TT) lines drawn tangential to the production possibility frontiers (PPF). In each diagram PPF1 is identical and the initial position has production at *a*, consumption at *b*, and *cb* units of the importable being obtained for *ac* units of the exportable. In Fig. 6.1a PPF2a is shown, representing a situation in which production possibilities have improved (say, because of technological change) in the exportable but not at all in the importable. With the terms of trade constant the results would be quite dramatic: trade would expand with *ef* units of the importable now being obtained for *ed* units of the exportable; production moves to *d*, with consumption at *f*. Note that at *f* real income is higher than before, and at *d* real output is higher than before but that production of the importable has contracted both relative to that of the exportable *and* absolutely.

Compare the situation in Fig. 6.1b. An alternative shift to PPF2b is shown where production possibilities have improved in *both* sectors sufficiently to again allow the same real income as with PPF2a with consumption at *f*. By comparison with Fig. 6.1a it is apparent that the volume of trade is lower and the composition of output is different, at *g*. In particular, this more balanced growth of production possibilities leads to a lower share of industry in total output *and* a lower absolute amount of industrial output.[7]

There is an intuitive argument which captures the flavour of the above for anyone who is uncomfortable with figure 6.1. It is advantageous to trade goods which can be produced relatively

[7] At *g* agriculture would be buying a higher share of industrial output than at *d*, yet agriculture's expansion has generated a lower level of industrial output!

cheaply at home for those which can be produced relatively cheaply abroad. When productivity grew very fast in Britain's exportable textiles, had productivity not increased at all in agriculture it would have been appropriate to specialize still more in textiles and import more food. Because agricultural efficiency actually improved the gains from more extreme specialization in textiles were less and it was cost effective to produce relatively more food at home.

In practice, for Britain the situation became more complicated by the early nineteenth century because by then Britain had a large share of world markets in major exports, and further expansion of exports tended to lower Britain's terms of trade. Davis very nicely indicated agriculture's role at this point: 'the importance of the increase of agricultural production at home (was that it) limited the need for imports of grain, and other foodstuffs, and so tended to keep down imports and prevent further depression of the terms of trade' (1979, p. 71). The terms of trade are discussed more fully in Chapter 7.

The example has been a simple one but it serves to illustrate an important point. In an open economy, like Britain during the industrial revolution, improving efficiency in the importable (agriculture) could *retard* the industrialization of output *ceteris paribus*. This point has been somewhat obscured in the literature, which has tended to assume the opposite. It should be noted that this argument is perfectly consistent with the discussion and evidence of Section II. There it was noted that in general agriculture had the *potential* to 'release labour' through increasing efficiency sufficient to permit industrialization with steady growth in a *closed* economy. It was also noted that in practice the economy grew faster than would have been consistent with agriculture's ability to release labour in a closed economy by having access to international trade. Under those circumstances agricultural efficiency increases served to limit the gains from international trade but by no means to eliminate them, as witness the very substantial amount of imported food in the early nineteenth century.

V CONCLUSIONS

The textbook literature on the industrial revolution has had much to say about the contributions of agriculture to industrialization. The discussion has often been in terms of agriculture's roles in industrialization, which have been thought of as potentially feeding a growing and increasingly urban population, releasing labour for employment in the agricultural sector, transmitting a flow of savings to the industrial sector, saving foreign exchange by obviating the need for food imports, and providing a growing market for industrial goods. The general assessment of the literature seems to be that agriculture largely succeeded in the first, and fourth, and the fifth roles, had some success in the third role, and failed in the second (Deane, 1979; Jones, 1974).

This chapter has also analysed the role agriculture played in structural change and has suggested rather different answers. It has also tried to indicate the value of distinguishing between the ex-post observable events and the counterfactual possibilities. Both are important in understanding structural change and agriculture's part in it.

A number of facts concerning intersectoral transactions have emerged in the preceding sections. The most important of these are as follows. First, Section II provided further evidence to suggest that British agriculture was very efficient by international standards and that capital accumulation plus technological change were important in raising output per worker. Second, Section IV illustrated that net food imports, which were insignificant in the first half of the eighteenth century, had risen to about one sixth by the early nineteenth century, and 22 per cent by 1841, and that in turn the economy had a substantial surplus in manufactures. Third, it became clear in Section IV that agriculture did not buy a very large share of industrial output in the eighteenth century—30 per cent seemed a reasonable guess—and bought less thereafter; but even so exports were responsible in the eighteenth century for rather less than the 40 per cent of the sales of industrial output. Fourth, as time went on the average output of an agricultural worker fed more urban workers. Combined with the data for agriculture's share in the labour force in Chapter 2, the evidence of Section IV suggests that each agricultural worker's output was capable of feeding 2.7 non-agricultural workers in 1841 compared with 1 in 1760. At the same time, earlier chapters have shown that the agricultural labour force grew in absolute numbers from 1760 to 1851, although only quite slowly. Fifth, we saw in Section III that it is probable that agriculture did generate a flow of savings towards the rest of the economy, although even by the 1820s this was probably able to finance at most a fifth of the non-agricultural investment which was taking place. These results clearly modify the findings of the traditional literature by suggesting that imports played quite a large part in food supplies from the early nineteenth century on and in agreeing with O'Brien that there has been a tendency to exaggerate the importance of agricultural markets for industrial goods.

Further changes to the traditional literature are also suggested by the counterfactuals considered during the chapter. These modifications are rather more radical and in effect examine agriculture's contributions by looking at the potential flexibility of the economy. The main points are as follows. First, the discussion of Table 6.5 indicated that for a given level of output a lower level of trade would be expected to reduce industry's share of output and the labour force; such an effect might have happened had policies been geared towards still higher levels of protectionism. Second, a more balanced expansion of production possibilities tends to reduce trade and also to lead to less change in the structure of output. Thus, in the context of an open British economy, the agricultural improvements led to a slower change in the composition of output than if the same overall productivity growth had been concentrated in exportables. Third, Section II established that agriculture was capable of releasing labour in the sense that, except for the late eighteenth century, it was possible to have steady growth of per capita output together with industrialization of the labour force without agricultural price rises in a closed economy. Table 6.5 suggests that this was a major reason for Britain's unusually pronounced industrialization of employment by 1840.

It seems likely that in the near future more counterfactual propositions will be investigated in the literature. This is to be desired since such analyses are important in understanding structural change. Perhaps footnote 7 of this chapter gave a particularly clear indication of this point. That footnote pointed out that in the model of Figure 6.1 an agricultural expansion which reduced the rate of structural change in the composition of output would *after the fact* show a *higher* share of industrial output purchased by agriculture.

Two final points are worth remembering. First, it seems clear that by European standards British agriculture had a high level of efficiency in the early nineteenth century even though agricultural goods were importables. The implication from trade theory is straightforward; namely, that relative efficiency in exportables (a subset of manufactures essentially) must have been still higher. The basis of this comparative advantage and its effects on real incomes in Britain are the main theme of the next chapter. Secondly, this chapter has in effect accounted for the unusually early equalization of the shares of agriculture in the labour force and in output which was stressed in Chapter 3. Agricultural improvement, which was particularly marked in Britain, helped to maintain the share of agriculture in output while also reducing its share in the labour force.

7

Specialization and International Trade

I INTRODUCTION

In this chapter a more detailed view of Britain's position in international trade during the 'Industrial Revolution' is presented. Previous chapters have already indicated that the openness of the British economy was importantly related to the pattern of structural change which was experienced, and this point is explored more fully in Section II. The evidence is familiar and comes from the well known works of Davis (1969) (1979) and Deane and Cole (1962). The picture which emerges is one of exports comprising a large part of the output of some sectors (notably in textiles and iron), and also that those same sectors accounted for a very high proportion of British exports (still about ⅔ in 1870).

The concentrated structure of Britain's exports immediately prompts the question, what were the sources of comparative advantage in the British economy? Since the notion of comparative advantage is that exports come from relatively more efficient sectors among tradable goods, and since relative efficiency is usually thought to be related at least partly to the availability of factors of production, examination of the pattern of trade can reveal something about the productive abilities of the home economy. This is undertaken in Section III. The results obtained there connect to the discussion of Chapter 4 on productivity growth and modernization during the industrial revolution, and together with the pattern of trade described in Section II indicate both that the much-used phrase 'Workshop of the World' is a misleading description of mid-nineteenth-century Britain and that the economy's trading base was built on coal and unskilled labour rather than large accumulations of human capital.

The rapid development of the few key exports was associated, particularly with cotton, with a tendency for export prices to decline during much of the industrial revolution. Section IV considers changes in the terms of trade between exports and imports in some detail since this is an important aspect of the extent to which welfare gains could be had from the uneven increases in productive potential. In common with other writers (for example, Hartwell and Engerman, 1975) the conclusion reached is that trade was not immiserizing. In fact, growth of demand abroad is undoubtedly of significance in this regard, that is, in maintaining real income levels, even for those economists who prefer to see the industrialization and growth processes as supply determined.

II THE PATTERN OF TRADE

In Chapter 6 it became apparent that the basic pattern of British trade during the industrial revolution was an export of manufactures in exchange for imports of food and raw materials. Confirmation and fuller details are shown in Table 7.1. It should be noted that the changes between the mid-eighteenth and mid-nineteenth centuries in terms of these broad categories are relatively minor, and that the industrial revolution saw the consolidation of existing tendencies.

The evidence of Table 7.1 complements the discussion of Chapter 3 which established that no other European economy had its exports so heavily in manufactures so early on or at such a low income level. In fact, Table 3.5 reported that France and Germany in 1870 still only had just over half their exports in manufactures. Together with the results of Table 6.7, which showed that at times during the industrial revolution as much as 35 per cent of industrial output was exported, Table 7.1 shows why mid-nineteenth-century Britain has been seen as the 'workshop of the world' and industrially supreme.

It is important, however, to put this 'supremacy' into perspective, and this is done by examining a more disaggregated picture. Table 7.2 shows that British merchandise exports were heavily concentrated in a few commodities, especially textiles with first woollens and then cottons as the major export sector. In addition, Table 7.3 demonstrates that cotton, in particular, sold a very high proportion of its output abroad, and that the woollen and iron industries also sold quite substantial parts of their output as exports.

Table 7.1. Composition of British Trade (%)

	Manufactures	*Raw Materials*	*Foodstuffs*
(i) Exports			
1700	80.8	8.2	11.0
1750	75.4	16.8	7.8
1801	88.1	5.0	6.9
1831	91.1	5.5	3.4
1851	81.1	13.3	5.6
(ii) Retained Imports			
1700	28.4	45.0	26.6
1750	14.4	54.5	31.1
1801	4.9	56.2	38.6
1831	2.2	70.4	27.4
1851	4.9	58.2	36.9

Sources: Derived from Davis (1969) and (1979), using adjacent years and accepting Davis's categorization. Note that as a result there would be a difference from the figure for the proportion of exports which were manufactures taken from Kuznets (1967) for Table 3.5. The proportions are based on current values.

Table 7.2. Shares of Major Exports in Total Merchandise Exports (%)

	Cottons	*Woollens*	*Iron and Steel*
1700	0.5	68.7	
1750	1.0	46.7	
1801	39.6	16.5	9.3
1831	50.8	12.7	10.2
1851	39.6	14.1	12.3

Sources: 1700 and 1750 derived from Davis (1969); 1801 based on figures for 1804/6 in Davis (1979), and 1831 and 1851 from Deane and Cole (1962, p. 31).

Table 7.3. Proportions of Gross Output Sold Abroad (%)

	Cotton	Wool	Iron
1760	50	46	
1801	62	35	24
1831	56	19	23
1851	61	25	39

Sources: Deane and Cole (1962, p. 185, 187, 196, 225) and for wool in 1760 based on Deane (1957, p. 215) for 1772.

There is, however, an important corollary of Tables 7.2 and 7.3, namely that the proportion of output exported in the rest of industry was very small. Thus, for example, in industries other than textiles and iron the same sources together with Table 6.7 show that the percentage of gross output exported in 1831 was only 7.7 per cent, and in 1851 5.8 per cent. Similarly, in 1700 of industrial output other than woollens only 7.5 per cent was exported. Throughout the industrial revolution there were sectors of manufacturing which were virtually non-traded internationally; for example, building and baking.

For much of the nineteenth century international trade in manufactures was heavily concentrated in textiles, presumably largely because in other industrial sectors transport costs were high relative to differences in comparative advantage. In 1880, textiles and clothing were 55.7 per cent of world trade in manufactures by value (Saul, 1965, p. 13). In that sector Britain was still in 1880 responsible for 46.3 per cent of world exports: in cotton alone perhaps 80 per cent (Sandberg, 1974, p. 141).

It is well recognized in the literature on international trading relationships for Britain during this period that the pattern of exports by commodity which emerged, following the productivity growth in textiles, brought with it a distinctive but changing geographic pattern to exporting. The economy of the eighteenth century exported its manufactures to Europe and increasingly to America. By the mid-nineteenth century areas other than these absorbed 43 per cent of British exports. The change in cottons was still more pronounced. Whereas in 1804/6 Europe and America took 92 per cent of exports, by 1854/6 their share was only 45.7 per cent, less than the 50.1 per cent going to Asia, the Near East, and Latin America. Whilst Britain dominated world exports, by the mid-nineteenth century, in the richer countries of Europe and also in the United States, home production had to a very large extent replaced imports from Britain, and the most important single market for British cotton exports was India (Farnie, 1979, pp. 90–101).

To a very considerable extent, then, Britain's very high share of world trade was based on dominance of textiles and, in particular, cotton, rather than a massive broadly-based comparative advantage in most spheres of manufacturing. This is not surprising nor necessarily a reason to criticize the mid-nineteenth-century economy. It

Table 7.4. Geographical Distribution of Manufactured Exports (%)

	Europe	*USA, Canada, West Indies*	*Africa, Near East, Asia, Australia, Latin America*
(i) All Manufactures			
1699/1701	83.6	13.3	3.1
1772/4	45.0	46.9	8.1
1804/6	37.3	49.4	13.3
1834/6	36.3	34.7	29.0
1854/6	28.9	28.1	43.0
(ii) Cottons			
1699/1701	20.0	80.0	0.0
1772/4	20.4	79.6	0.0
1804/6	47.1	45.1	7.8
1834/6	47.4	19.8	32.8
1854/6	29.4	16.3	54.3

Source: Derived from Davis (1969) (1979) based on values in current prices, except for 1772/4 which are based on official values.

is, however, worth noting that this pattern of comparative advantage as revealed by the trade figures goes along with the pattern of modernization and productivity growth by manufacturing sector revealed in Chapter 4. Exports did not come to a great extent from the unmodernized traditional sectors which, as Musson and Clapham reminded us, still constituted much of the economy.

III COMPARATIVE ADVANTAGE IN MID-NINETEENTH-CENTURY BRITISH MANUFACTURING TRADE

The notion of comparative advantage is simple and widely understood, and was clearly stated by Ricardo (1817). The idea is that a country will tend to export commodities which in the absence of trade it would produce relatively efficiently, that is, at a lower opportunity cost than the rest of the world, and to import those goods which it would produce at a relatively high opportunity cost. The basic argument was extended by Heckscher-Ohlin who predicted that a country would tend to have a comparative advantage in those goods which used intensively the factor of production in which the country was relatively abundant (Caves and Jones, 1981, ch. 7). In a modified form this approach yields a powerful description of international trade where 'inter-country differences in the structure of exports are largely explained by differences in physical and human capital endowments' (Balassa, 1981, p. xx). Balassa goes on to suggest that in this scheme of things in the mid-twentieth century the most advanced country has its comparative advantage based on intensive use of human capital and is relatively efficient in human capital-intensive goods. This hypothesis is borne out by Stern and Maskus's (1981) results, together with a finding that the United States also has a relative superiority in goods of recent origin for which research and development and employment of scientific personnel are important inputs.

By contrast, Britain a century earlier appears to have occupied a quite different position in international trade. Britain's comparative advantage in manufacturing in the third quarter of the nineteenth century seems to a major extent to have been rooted in coal and unskilled labour. Textiles and iron were the major users of steam power (Kanefsky, 1979), and cotton textiles in particular relied heavily on low-grade labour (Farnie, 1979, pp. 177–8; Sandberg, 1974, pp. 214–15).

This was clearly perceived by Victorian economists. Ashley stressed the substantial export of 'commodities produced by poorly-paid and low-grade labour in our large towns' (1904, p. 6), whilst Jevons, writing in the 1860s, saw trade as follows: 'Great Britain, capable for the present of indefinitely producing all products depending on the use of coal [and] Continental Europe, capable of an indefinite production of artistic, luxurious, or semi-tropical products, but debarred by comparative want of coal from competition with us' (1965, p. 416).

Indeed earlier chapters have prepared us for this picture. Chapter 3 noted Britain's extremely high coal consumption. Chapters 3 and 4 established that Britain's economic development was characterized not by a manufacturing sector with generally high value-added per worker, but by having an exceptionally low proportion of its labour force remaining in agriculture at any given income level. The use of much unskilled labour goes with the rather low level of human capital formation discussed in Chapter 3; as Sanderson notes, during the industrial revolution 'the new technology could operate with an illiterate labour force which it had helped to produce, and economic growth was not impeded by educational retardation' (1983, p. 16).

An application of Stern and Maskus's model to 1880 supports this account of British exporting. Exports are found to be higher in sectors which are horse-power or unskilled-labour intensive, and lower in sectors which are intensive in the use of human capital (Crafts and Thomas, 1984, p. 13).

IV THE TERMS OF TRADE

It was noted in Chapter 6 that the gains from international trade depend on the terms of trade; that is, the price of exports relative to imports. A striking feature of Britain's nineteenth-century growth is that during the first half of the century the price of exports fell sharply relative to imports. The evidence is given in Table 7.5.

If the source of changing terms of trade is an exogenous change elsewhere in the world, for example changing weather conditions affecting the prices of imported foodstuffs, then the interpretation of the situation is straightforward: falling terms of trade would represent a reduction in real income as the economy's purchasing power in terms of imports declined. In the case of Britain after 1800, however, the position is rather different because we know that the major factor influencing the terms of trade was the fall in cotton-goods prices resulting from productivity growth. The export price of cotton yarn (40's), for instance, was 16/– in 1779, 7/6*d*. in 1799, 2/6*d*. in 1812, 1/2½*d*. in 1830, and 0/11½*d* in 1860 (Ellison, 1886, p. 61).

Table 7.5. The Net Barter Terms of Trade (1880 = 100)

1801	196
1811	169
1821	154
1831	136
1841	109
1851	108

Source: Imlah (1958, pp. 94–6); the figures are 5-year averages centred on the date shown.

The declining terms of trade were then a consequence of expanding productive potential in Britain. It is a possibility well recognized in the economics literature for increased productive potential actually to reduce real income by depressing the price of exports so much relative to imports that the gains in output are swamped by the loss of purchasing power for imports. In any case, if growth at home tends to depress the terms of trade, this will impose an important restraint on the extent to which extra output can be converted into extra real income. This would be most likely to be significant for a country in which growth of productive potential was concentrated heavily in a few exportables and which had a large world-market share in those goods.

We saw earlier in this chapter that Britain's exports were heavily concentrated in a few sectors, and also, in Chapter 4, that productivity growth in manufacturing was skewed towards these staple exports. Despite these tendencies British growth was not immiserizing and it is interesting to see why this was so. Important factors preventing immiserization and, by the second quarter of the nineteenth century, permitting relatively rapid expansion of home output without immiserization, were productivity improvements in agriculture and growth of export demand as incomes abroad increased.

In order to explore the effects of increased productive potential on economic welfare it is necessary to use a more refined concept of the terms of trade, the *single factorial* terms of trade. In principle this measure weighs the extra productivity of factors of production in exportables against the decline in the net barter terms of trade weighted by the share of imports in home consumption (that is, the reduction in real income from lower import-purchasing power). In practice, the approximation usually used considers growth of labour productivity net of extra capital formation (Spraos, 1983, pp. 70–4). Decline of the single factorial terms of trade indicates immiserizing growth. Thus we have

$$\frac{\Delta\,\text{SFTT}}{\text{SFTT}} = w.\frac{\Delta\text{NBTT}}{\text{NBTT}} + \frac{\Delta(Y/L)}{Y/L} \qquad (7.1)$$

where SFTT is the single factorial terms of trade, w is the share of imports in home consumption, NBTT is the net barter terms of trade, and Y/L is (net) labour productivity in exportables. In early nineteenth-century Britain w was about 0.2.

Table 7.6 reports the behaviour of (7.1) for 1801–1851 using the revised estimates of growth in industrial output per worker to approximate $\Delta(Y/L)/(Y/L)$. Table 7.6 shows that over 1801–1831 the weighted single factorial terms of trade grew only very slowly, although after 1821 growth of labour productivity outstripped the decline in the net barter terms of trade. In effect, Table 7.7 suggests that Britain had virtually reached the limits to growth based on the prevailing composition of productivity increase and international trade in the early nineteenth century. Nevertheless, in the long run, as Hartwell and Engerman (1975) concluded, growth was not immiserizing.[1]

We can now gain further insight into the avoidance of immiserization with the aid of simple formulae used in international economics. Derivations of (7.2) and (7.3) are shown in the appendix to this chapter, based on Caves and Jones (1981, pp. 496–9).

An expansion of home output of the exportable tends to depress the terms of trade, the more so if the extra income is spent on importables and the more so if prices have to move substantially to adjust the balance of payments because import demands are not

Table 7.6. Rate of Change of Single Factorial Terms of Trade (% per year)

	$\Delta SFTT/SFTT$	$\Delta NBTT/NBTT$	$\Delta(Y/L)/(Y/L)$
1801–51	0.63	−1.20	0.87
1801–31	0.04	−1.23	0.29
1821–51	1.33	−1.19	1.57

Source: derived from Table 7.5 and the estimates of Chapters 2 and 4. Y/L was adjusted for changes in industrial capital formation based on Feinstein (1978, p. 41). Throughout w was taken to be 0.2.

very responsive to prices. The change in real income resulting for a change in the output of the exportable is

$$\Delta y = \left(\frac{e + e^* - 1 - m}{e + e^* - 1}\right)\Delta q_x \tag{7.2}$$

where y is real income, e (e^*) is the home (rest of world) price elasticity of demand for imports, m is the marginal propensity to import, and q_x is output of the exportable.

[1] Hartwell and Engerman's conclusion was based on Deane and Cole's over-optimistic estimates of labour-productivity growth, and implicitly assumed $w = 1$ by failing to allow for the fact that not all consumption was imported. It should be noted that my revised figures for industrial labour-productivity growth would give *declines* in the single factorial terms of trade if w is taken to be 1.

The value of m during 1800–1850 was probably about 0.2, and fairly constant to judge from Davis (1979, p. 86) and Deane and Cole (1962, p. 166). The values of e and e^* are not known for this period; in the early twentieth century they were probably about 0.5 and 1.5 (Moggridge, 1972, p. 245). Given Britain's much greater dominance of trade in manufactures in 1850, and the lower share of manufactures in her imports, it is quite likely that the elasticities were lower in 1800–1850.

This immediately reveals the role of agriculture in supporting the terms of trade, the argument made by Davis (1979, p. 71). With an income elasticity of demand for food of 0.7, and about 35 per cent of total expenditure on food, the marginal propensity to spend on food was about 0.25, whereas the evidence of Table 7.1 suggests that the marginal propensity to import food was only about 0.05. Thus, in the absence of agricultural expansion the overall marginal import propensity, m, might have risen to around 0.4, a value which conceivably could have implied immiserizing growth and would certainly have worsened the terms of trade.

More generally, changes in real income occur either through rises in production or imports becoming dearer or cheaper (the terms of trade effect); the magnitude of this effect depends both on how important imports are and the size of the change in the terms of trade. The terms of trade will tend to rise when demand for British goods expands abroad, and to fall when British demand for imports expands. The impact of such shifts in demand depends on how much prices have to adjust to restore equilibrium in the balance of payments. Thus the growth of real incomes is

$$\frac{\Delta y}{y} = \left(\frac{e + e^* - 1 - \theta m}{e + e^* - 1}\right)\mu + \frac{\theta m . \mu^*}{(e + e^* - 1)} \qquad (7.3)$$

where $\mu(\mu^*)$ is the home (rest of world) rate of growth of demand for imports at constant prices, and θm is the share of imports in national income. Note that shifts outward of demand abroad raise real income.

Equation (7.3) helps to establish the second argument made above, namely that the retention of welfare gains in Britain was aided by faster growth abroad. Maddison (1983, p. 45) notes that after the Napoleonic Wars world growth accelerated; the 'soon to be developed' countries grew at about 2 per cent per year between 1820 and 1870 compared with well below one per cent in the eighteenth century. Britain's share of world trade remained fairly constant (Mulhall, 1884, p. 128). A crude approximation might then put μ^* at 2 per cent after 1820, $(e + e^* - 1)$ at 0.5, and θm at 0.2; this would give a value for the last term in equation (7.3) of 0.8 per cent, quite sizeable by comparison with overall growth rates and with the eighteenth century.

Thus, whilst the period after the Napoleonic Wars may not be export-led growth, it is worth recognizing that real income growth was supported by export-demand growth and that this was a much more favourable element in the economic environment than earlier.

V CONCLUSIONS

This chapter has chiefly been a refinement and extension of themes already to be found in earlier parts of the book. To a certain extent it has been speculative, in that research on the aspects of trade considered here is still in its infancy. Nevertheless, it is important to extend our appreciation of the role of the external sector beyond the simplistic claims of export-led growth which we encountered in Chapter 6.

The description of Britain's trade given in Section II helps us further to understand the idiosyncrasy of British economic development. Without its export market the cotton-textile industry would obviously have been much smaller, as indeed would the 'revolutionized' as opposed to the traditional sector of manufacturing. The technological lead which Britain briefly enjoyed in textiles, which were tradables, to a considerable extent accounts for the greater degree of industrialization based on trade found in Table 6.5. The expansion of trade for the export-oriented cotton industry is clearly fundamental to the 'factory–steam' definition of 'Industrial Revolution' which figures so much in the historiography and which was much more characteristic of British than, say, French development.

Section IV provides further evidence of agriculture's important part in supporting the terms of trade, and confirms that this contribution of agriculture has been unduly neglected by most earlier writers, as Chapter 6 indicated. Even so, Section IV showed that there was a considerable fragility about the process of real income growth until growth spread more generally in the rest of the world. The fragility is reflected in the weak performance of the weighted single factorial terms of trade before the 1820s, and seems to indicate that the concentrated pattern of productivity growth and export expansion could not have proceeded much faster in these early stages without exhausting the scope for welfare gains.

This theme of impressive productivity improvement in a relatively few sectors which was brought out in Chapter 4 seems also to have its counterpart in Britain's revealed comparative advantage. Whilst Section II showed that exports were heavily concentrated in a few commodities, Section III brought out that comparative advantage for the economy overall in the so-called 'workshop of the world' era was to be found in goods intensive in the use of coal and unskilled labour. The relative disadvantage in human-capital intensive goods reinforces several points made in earlier chapters. It might be expected of an economy which was *not* pervasively innovative. It probably reinforces the argument that the technological lead Britain built up in the early industrial revolution was not inevitable, and it is entirely consistent with the theme of Chapter 3 that Britain's structural transformation was heavily orientated towards the establishment of a high share of the labour force in manufacturing rather than high output per worker within industry.

In fact, Section III relates some, relatively little-noted, manifestations of Britain's early start. Understanding the basis of Britain's comparative advantage, and the 'peculiar' development process the economy experienced during the industrial revolution, is of considerable importance for an appreciation of the debate over late-Victorian economic failure and even the traumas of the interwar economy. It is to these topics that the final chapter is addressed.

APPENDIX TO CHAPTER 7

The derivation of the formulae (7.2) and (7.3) is tedious but not difficult. Let p be the relative price of the home country's imported good and note that $dp > 0$ represents a decline in the net barter terms of trade. Asterisks again represent the rest of the world.

Balance of payments equilibrium is

$$pM = M^* \qquad (7.4)$$

where M is imports. Therefore

$$\hat{p} + \hat{M} = \hat{M}^* \qquad (7.5)$$

where ^ denotes a proportional rate of change. Imports in either country respond to changes in the terms of trade and to changes such as income growth which raise imports at constant prices

$$M = -e\,\hat{p} + \hat{M}|_{\bar{p}} \qquad (7.6)$$

$$\hat{M}^* = e^*\hat{p} + \hat{M}^*|_{\bar{p}} \qquad (7.7)$$

Combining this with (7.5) gives an expression for the growth rate of the net barter terms of trade for shifts in import demand

$$\hat{p} = \frac{\hat{M}|_{\bar{p}} - \hat{M}^*|_{\bar{p}}}{e + e^* - 1} \qquad (7.8)$$

Changes in real income can be decomposed into a terms of trade effect and a direct effect from production changes (dq)

$$dy = -\,M.dp + dq \qquad (7.9)$$

or multiplying the first term top and bottom by p

$$dy = -pM.\hat{p} + dq \qquad (7.10)$$

Consider a growth of home output at a rate μ which involves import growth *pari passu* (roughly characteristic of early nineteenth-century Britain) and divide (7.10) throughout by y

$$\hat{y} = -\theta m\hat{p} + \mu \qquad (7.11)$$

Combining this with (7.10) and assuming $\hat{M}^* = 0$ gives

$$\hat{y} = \left(\frac{e + e^* - 1 - \theta m}{e + e^* - 1}\right)\mu \qquad (7.12)$$

If there were only growth abroad at a rate of μ^*, then using (7.8) and (7.11) gives

$$\hat{y} = \frac{\theta m.\mu^*}{e + e^* - 1} \tag{7.13}$$

Combining (7.12) and (7.13) gives (7.3) (QED).

Consider now expansion of the exportable q_x. The associated increase in demand for importables will be

$$dM = \frac{m.dq_x}{p} \tag{7.14}$$

where m is the marginal propensity to import. Using (7.8) this implies that

$$p = \left[\frac{m}{pM\,(e + e^* - 1)}\right] dq_x \tag{7.15}$$

Now it is straightforward to use (7.10) to obtain

$$dy = -\left[\frac{m}{e + e^* - 1}\right] dq_x + dq_x \tag{7.16}$$

which can be rearranged to obtain (7.2) (QED).

Finally, it was noted in the text that $(e + e^*) > 1$. This is simply the famous Marshall–Lerner condition for balance of payments stability.

8

Some Legacies of the Early Start

I INTRODUCTION

THE preceding chapters have drawn a picture of the industrialization process in Britain that is in several respects substantially different from the conventional wisdom that prevailed until recently. If the account given is broadly correct, then implications can be drawn out for our view of later developments in Britain's economic history, and areas where further research may be fruitful can be suggested. This is the task of this final chapter, which necessarily is more speculative than the rest of the book and which should be regarded as provocative rather than definitive.

There are two areas of considerable controversy concerning later periods for which insights can be found by following up the lines of argument developed in earlier chapters, namely the discussion of declining economic growth in the late nineteenth and early twentieth centuries, and the debate over the contribution of structural adjustments to the high unemployment which followed World War I. Both these debates are, of course, very well known to undergraduates who study recent British economic history, although they are often taught with fairly little reference to the literature on the 'Industrial Revolution'.

A number of themes have been developed earlier in the book which can usefully be recapitulated at this stage. In Chapter 2 I presented the data to show that for much of the industrial revolution period national product in Britain did not grow very rapidly. When in Chapter 4 these data were used to examine productivity growth, it emerged that productivity growth was concentrated in a few well-known sectors of the economy, and that indeed manufacturing-productivity growth as a whole was very low until the second quarter of the nineteenth century. By contrast, Chapter 3 established both that structural change was very rapid and can indeed be thought of as an 'Industrial Revolution' in the century or so before 1850, and also that in several ways the path that Britain took to economic development was very different from the 'European Norm'. Notably, Britain experienced relatively low levels of accumulation of both human and physical capital and redeployed a remarkable proportion of the labour force out of the primary sector and into manufacturing. On the other hand, it was found that Britain did not attain particularly high levels of output per worker in industry, and as a useful crude approximation it can be said that the triumph of the industrial revolution lay in getting a lot of workers into industry rather than obtaining high productivity from them once there. Finally, in Chapter 7 it was established that Britain developed with a pronounced pattern of specialization in exports, relying heavily on the staple industries, and arrived at the end of the nineteenth century with a comparative advantage in low rather than high-wage sectors.

Bearing these points in mind we can now turn to the two debates in question. As when in Chapter 5 the standard of living debate was discussed, the summaries of the literature to be given are, of course, brief, and for those unfamiliar with the historiography it will be necessary also to consult a full textbook account.

II DID VICTORIAN BRITAIN FAIL?

During the late 1960s and early 1970s there was a strong move by neo-classical economic historians to defend the performance of the British economy in the late nineteenth and early twentieth centuries against the criticisms of an earlier group of writers. The attempt at exoneration was spearheaded by McCloskey, the title of whose best-known paper heads this section of the chapter. In that paper McCloskey painted 'a picture of an economy not stagnating but growing as rapidly as permitted by the growth of its resources and the effective exploitation of the available technology' (1970, p. 451). McCloskey attempted to show that at the macro-economic level productivity growth was creditable compared with other advanced economies (1970, p. 458), and that there was no substantial gain to be had in redirecting foreign investment to the domestic economy (1970, p. 455). This last point was supported by the work of Edelstein (1976), who indicated that the London capital market generally allocated funds efficiently and did not exhibit a tendency to invest too much abroad. A number of neo-classical studies of entrepreneurial decision-making supported the argument that productivity performance was respectable by obtaining results that suggested that, in general, entrepreneurs adopted the right technology for British conditions, taking into account especially relative factor costs (Sandberg, 1981). Finally, with regard to Britain's trade, Harley and McCloskey argued that Britain was right to continue all the way to World War I to exploit her comparative advantage in the old industries, and that her relative slowness to expand new industries was appropriate given her skills and factor endowments (1981, pp. 68–9). Thus, if growth was slower than elsewhere this should be seen as the result of unavoidable supply constraints and other countries exploiting a backlog of technological opportunities already adopted by Britain, if profitable (McCloskey, 1970, p. 451).

Yet this neo-classical defence of the economy has not found universal support even among economic historians who base their arguments on economic analysis. For example, significant details of entrepreneurial performance and institutions in the important and much-discussed steel industry have once again been criticized recently (Allen, 1981; Berck, 1978; Webb, 1980), and Allen (1979) has demonstrated that total factor productivity in the industry in Edwardian times was lower than in either Germany or the United States. Matthews *et al.* (1982) have argued that total factor-productivity growth was much slower than the data previously available had suggested and was accordingly slower than in other advanced countries. Kennedy took the view 'not that British resources were incapable of sustaining more rapid growth, but rather that resources were not deployed to exploit opportunities which did exist' (1974, p. 440), and in subsequent papers pointed both to deficiencies in the capital market concerning new industrial issues (1976) and to the structure of Edwardian Britain being tilted away from modern, technologically-orientated manufacturing when compared with that of the United States (1982).

This last argument is reminiscent of an earlier thesis put forward by Richardson, which has been popular with historians ever since. Richardson argued that in the years before World War I the legacy of the early start was that the economy was 'overcommitted' and that 'The hindsight argument is that a more rapid transfer of resources to the new industries should have been achieved earlier' (1969, p. 194). Richardson was particularly concerned with 'the degree of elasticity in the supply of factors of production and the complementary question of the mobility of these factors under the influence of long term growth trends in declining and expanding industries' (1969, p. 196), and he argued that the early start had produced an entrepreneurial psychology and an institutional structure which retarded structural change (1969, p. 195). In his turn, Richardson followed a long tradition of writers seeking to argue that Britain's early start in industrialization adversely affected later growth performance.

III THE EARLY START, THE CLIMACTERIC, AND OVER COMMITMENT

In Chapter 4 a growth-accounting exercise was undertaken in order to examine the contribution of total factor-productivity growth during the industrial revolution. The data were admittedly crude, but nevertheless it seems useful to recall these results and compare them with the estimates obtained by Matthews *et al.* with better data for the later years of the nineteenth century.

As is by now well-known, Matthews *et al.* date the chief slowing-down in growth of output and productivity in the period 1899–1913. Certainly for the economy as a whole or manufacturing this seems appropriate. For mining (a subset of industry) and for agriculture the slowdown in productivity growth came in the 1870s, and the decline in the latter actually accounts for about half the total decline in productivity growth between 1856–73 and 1873–1913 recorded by Matthews *et al.* (1982, p. 229). Taken together with the results of Chapter 4 repeated in Table 8.1, this serves to underline the importance of the rate of technical improvement in agriculture as a determinant of overall productivity growth, a point often forgotten in discussions of changing nineteenth-century growth rates. Indeed, as Feinstein *et al.* argue, 'the falling off in the rate of growth of total factor productivity in manufacturing was virtually confined to the period after 1899' (1983, p. 183). The best guess now seems to be that the long period from the end of the French Wars to the end of the nineteenth century appears as an epoch of a steady trend-growth in productivity in manufacturing.

The dating of the general climacteric in the period after 1899 should, however, probably be regarded as cold comfort and rather less convincing as a defence of British performance than McCloskey

Table 8.1. Total Factor. Productivity Growth, 1760–1913 (% per year)

	Whole Economy	*Industry*	*Agriculture*
1760–1800	0.2	0.2	0.2
1801–31	0.7	0.3	0.9
1831–60	1.0	0.8	1.0
1856–73	0.8	1.0	0.9
1873–99	0.7	0.7	0.5
1899–1913	0.0	0.0	0.4

Sources: the first three rows are based on Chapter 4, and the last three rows on Matthews *et al.* (1982), Tables 8.1 and L1. 'Industry' subsumes manufacturing, mining, and construction. The TFP figures from Matthews *et al.* are the crude estimates which do not allow for changes in the quality of the labour force.

(1970) imagined. For it is at this juncture that a cluster of innovations, sometimes called the 'second industrial revolution' (Landes, 1969, p. 235), based on a much greater use of applied science and research came to exert a strong influence on productivity growth and raised the rate of return to investment in human capital. Moreover, it was at this moment that total factor-productivity growth rose sharply in the United States from 0.35 per cent per year prior to 1890 to 0.84 per cent in 1890–1905, and 1.52 per cent in 1905–27 (Abramovitz and David, 1973). Similarly, in Germany total factor-productivity growth in the economy overall fell only slightly in 1900–13 (1.06 per cent) as compared with 1870–1900 (1.17 per cent), and in industry and commerce rose from 1.13 per cent in 1870–1900 to 1.46 per cent in 1900–13 (Andre, 1971, p. 124).

Thus the climacteric in productivity growth was not a general experience among advanced countries, even those like Germany and the United States both of whom had higher levels of output per worker in industry than Britain did by the first decade of this century. Indeed, Abramovitz and David see the turn of the century as the period when returns to investment in human capital could potentially raise the rate of growth much more significantly than had ever been the case earlier on. A consistent complaint of the critics of Britain's growth performance in this period, since Alfred Marshall, has been that Britain under-invested in education especially scientific and technical education.

This brief review of the evidence on productivity growth seriously undermines McCloskey's argument that growth was as fast as permitted by exogenously given supply constraints. It suggests that Britain was unable to take advantage of the possibility of a higher growth path based on the opening-up of higher returns to investment in education and science. This argument has not been investigated by neo-classical economic historians and is an important part of the 'overcommitment' hypothesis.

The work of Chapter 4 helps to put the productivity growth experience of the late nineteenth century into perspective. From the revised growth estimates it now appears that at no time during the industrial revolution did Britain attain a rate of total factor-productivity growth close to 1.5 per cent per year, and rapid productivity growth was by no means pervasively spread throughout the economy. Perhaps the right way to think of 'Edwardian failure' is in terms of the economy not being able to accelerate its productivity growth when the chance came.

If there was a potential for faster growth, why was it not realized? Recent research, together with the evidence of earlier chapters of this book, significantly clarifies the possibilities, although it certainly does not give a definitive answer. It no longer seems plausible to regard entrepreneurs as grossly inept in their decision-making, and the London capital market's allocation of funds shows signs of efficiency. As Saul has noted, crude assertions that the early start inevitably hampered later performance have not been shown to be convincing (1968, pp. 44–50).

Crafts and Thomas (1984) found that Britain's comparative advantage in goods intensive in the use of unskilled labour and a comparative disadvantage in goods intensive in the use of human capital persisted through to the 1930s, and this may seem consistent with Harley and McCloskey's claim that Britain was right to stay heavily in the old industries before World War I. In fact, it is possible to throw more light on this point by looking at a measure of 'revealed comparative advantage' where data allow a wider range of comparisons. The measure of revealed comparative advantage used is to compare a country's world-market share in each sector with its overall world-market share. This provides a ranking of relative export performance, and a revealed comparative advantage exists if a sector has a share greater than the average. Some results for 1913 and 1937 are shown in Table 8.2; a fuller set of results is available in Crafts (1985b).

The results of Table 8.2 confirm Harley and McCloskey's view that Britain had a very different comparative advantage from that of other advanced countries. In 1913 the 'old staples' dominate the rankings, whereas 'new' and technologically more sophisticated industries are more prominent in the rankings for Germany, the United States, and even France. Britain also had a rather different pattern of destinations for its exports, as is seen in Table 8.3, in which it is clear that the mid-century pattern of a high proportion of exports going to less advanced and semi-industrial countries was still the case in 1913. Coal, iron and steel, cottons, and woollens were still the main exports, accounting for 51.0 per cent of exports in 1913 compared with 64.6 per cent in 1851 and 62.2 per cent in 1873 (Mitchell and Deane, 1962, p. 305).

At this point, however, it is useful to consider the sources of comparative advantage. The view taken here, which was examined more fully in Chapter 7, is that comparative advantage is largely determined by factor endowments. This is important both because it indicates a way in which Britain's early industrialization process influenced later growth opportunities and because it implies that the comparative advantage of the economy in 1913 was the outcome in considerable part of earlier decisions about accumulation.

The comparison of development patterns in Chapter 3 indicated that Britain's pattern of capital accumulation was very different from the European norm; in particular, Britain had relatively low rates of capital formation and relatively high rates of foreign investment. The realization of the gains from technological progress in the industrial revolution does not seem to have depended on high investment in education (Sanderson, 1983) and foreign investment seems to have been profitable, at least in terms of private returns in the capital market of the day (Edelstein, 1976). Thus it seems that the pattern of British capital accumulation was idiosyncratic, but so were the opportunities available to the most heavily industrialized country in, say, 1870, and there is a strong chance that resources were being fairly efficiently used at the time. Nevertheless, this pattern of accumulation based on the early start bequeathed a stock of factor endowments to the late Victorian age which moulded comparative advantage at that time and thus helped to produce the structure of the economy.

This was a complicated process and is one which requires further research before links from the early start to later economic

Table 8.2. Revealed Comparative Advantage in Manufacturing

	United Kingdom	*Germany*	*United States*	*France*
1913	Rail and Ship	Electricals	Non-ferrous metals	Spirits/Tobacco
	Textiles	Cameras/Books	Agricultural Equipment	Motor Cars
	Iron and Steel	Leather/Wood	Industrial Equipment	Apparel
	Spirits/Tobacco	Industrial Equipment	Motor Cars	Cameras/Books
		Chemicals	Electricals	Finished Goods
		Metal Manufactures	Metal Manufactures	Leather/Wood
		Finished Goods	Leather/Wood	Textiles
		Iron and Steel	Rail and Ship	Chemicals
		Non-metal materials	Iron and Steel	
		Apparel	Cameras/Books	
1937	Spirits/Tobacco	Metal Manufactures	Agricultural Equipment	Spirits/Tobacco
	Textiles	Finished Goods	Motors/Aircraft	Apparel
	Rail and Ship	Chemicals	Industrial Equipment	Textiles
	Finished Goods	Cameras/Books	Electricals	Iron and Steel
	Electricals	Non-metal materials	Iron and Steel	Chemicals
		Rail and Ship	Non-ferrous metals	Non-metal materials
		Electricals	Cameras/Books	
		Industrial Equipment		

Source: Crafts (1985b), based on data from Tyszynski (1951). Sectors are placed in rank order.

Table 8.3. Destinations of Manufactured Exports in 1913 (%)

	Industrial	*Semi-industrial*	*Other*
United Kingdom	31.8	41.3	26.8
United States	63.2	16.2	20.6
Western Europe (inc. France, Germany)	58.1	13.2	28.7

Source: derived from Maizels (1963, p. 227).

structure can be properly understood. For example, Matthews *et al.* (1982, p. 526) have sensibly hypothesized that the economy may in the late nineteenth century have been vulnerable to 'Dutch disease'. The argument would be that with a large existing stock of assets abroad yielding a large positive contribution to the balance of payments manufacturing's international competitiveness was reduced below what would otherwise have been the case. In these circumstances a cessation of foreign investment by pushing up the terms of trade would only have made manufacturing imports more attractive and investment in home industry still less attractive and thus the economy was 'locked in' to foreign investment. A related point might be that this affected the perceived structure of rewards to investment in human capital, encouraging the production of 'City gents' rather than technologists. Certainly Rubinstein found that of millionaires dying in the period 1860–1919, 50 per cent more had fortunes based on commerce and finance than based on manufacturing (1981, pp. 62–3). Obviously, these are no more than tentative ideas which require considerable research, but they do suggest the beginnings of an intellectually respectable early-start hypothesis.

Nevertheless, arguments like this seem unlikely fully to answer the question as to why Britain failed to move to the higher growth path possible in the early twentieth century. There seem to be further features of the economy which served to reduce its flexibility and underwrote the pattern of comparative advantage revealed in Chapter 7 and Table 8.2. Three points stand out:

(1) Britain persistently had a relatively low rate of accumulation of human capital, and seems to have been exceptionally poor at producing scientists to work in industry. Although educational spending as a fraction of GDP rose, it fell further behind the

United States and Germany such that in 1900, while Britain spent 1.3 per cent of GDP on education, the US spent 1.7 per cent, and Germany 1.9 per cent. At the turn of the century, when it seems probable that returns to investment in human capital were rising, the rate of accumulation of skills per worker estimated by Williamson rose in the United States to 0.57 per cent per year, compared with 0.30 per cent in Britain (1981, p. 26). The educational system was much debated and could have been reformed substantially, yet, as Sanderson shows, 'By any criterion as a major industrial power our quantitative lag in producing scientists for industry was plain compared with the capability of our chief competitors' (1972a, p. 24). Britain's early start perhaps made it more likely that this would be so, and certainly the reforms of the mid-nineteenth century led to better clerical rather than scientific skills (Sanderson, 1983, pp. 29–47), but it did not make it inevitable that the shift towards more and new kinds of education would be so long delayed.

(2) In at least some new industries economies of scale and learning-by-doing seem to have been important aspects of comparative advantage. The best documented case is that of motor cars, where Foreman-Peck has shown the importance of both points (1979, 1982). In this case reversing an early failure to establish large-scale production in the industry seems to have required tariffs to protect and establish an industry which ultimately (in the 1950s) became one in which Britain had a substantial comparative advantage. Thus, it is possible that flexibility of policy response was required in the late nineteenth century whereas policy changes had little to do with Britain's success as an early industrializer.

(3) There are still some weaknesses of the capital market which require further investigation, notwithstanding the admirable work of Edelstein. In particular, Hannah (1974) has shown that the takeover bid did not operate effectively to eliminate poor management, and, in the absence of a well-developed market for industrial issues, much investment in industry appears to have been financed outside the London capital market or possibly to have been inadequately financed (Cottrell, 1980; Kennedy, 1976). Both these features of the capital market appear to stem largely from serious imperfections of information in the absence of either modern companies legislation or German-style integration of

finance and industry. These deficiencies may have held back earlier growth also, but perhaps they were more obvious and damaging in the circumstances of the 'second industrial revolution'.

These arguments together suggest that there may still be value in considering the British economy of the late nineteenth century to have been 'overcommitted' in Richardson's sense. The earlier chapters of this book offer some help in understanding how this may have happened. Nevertheless, it should be clear that the evidence of earlier chapters does not establish an automatic link from the early start to later relatively slow growth, although we have seen that Britain was an 'idiosyncratic industrializer'. Rather we have seen that the nature of Britain's early start seems to have *conditioned* the later ability to respond to new opportunities. The economy did eventually respond, however, and so did its comparative advantage, as the results for 1937 in Table 8.3 show.

IV THE NATURE OF INTERWAR UNEMPLOYMENT

The causes of the high level of unemployment in Britain between the wars have become a subject of renewed controversy recently. Economic historians have frequently argued that the problem was one of 'structural adjustment' to a large extent, involving a painful contraction of the old staple industries and the growth of newer industries in different locations.

In some views this adjustment problem appears as a natural consequence of the early start in industrialization, and the comparative advantage which ensued from it in the late nineteenth century but was eroded by the changing international position after 1914. Thus Harley and McCloskey argue that

> With the benefit of hindsight we can see that twentieth century Britain has paid a considerable price for the industrial concentration that led to the large exports of a few commodities to relatively few markets by the eve of the First World War. Unemployment in the interwar economy was concentrated in these old industries, because labour that had been drawn to these industries in the pre-war period could not be moved costlessly to other employments. The loss to the economy was enormous (1981, p. 69).

A variant on this argument is given by Richardson, who sees an ending of overcommitment in the interwar period which was 'a time of revitalization and rebirth to be contrasted with the pre-

1914 era' (1967, p. ix), and interprets high unemployment as a corollary of a catching up in productivity performance: 'A large proportion of the high unemployment of the 1930's could be described as technological unemployment. The persistence of large numbers out of work reflects the efficiency of recovery far more than its incompleteness' (1967, p. 23).

The degree of controversy among economists looking at the interwar unemployment problem has been greater, and concerns in particular the extent to which unemployment can be seen as a result of deficient aggregate demand, as Keynesians would argue, or as a result of a high level of unemployment even when the labour market was in equilibrium, a view more in keeping with a neo-classical model of the economy. Representing this school of thought Benjamin and Kochin (1979 and 1982) claimed that for most of the interwar period high unemployment should be regarded as voluntary and occasioned by unemployment benefits which were generous relative to wage rates. Their Keynesian critics, Hatton (1983) and Broadberry (1983), produce evidence to suggest that the economy was not in equilibrium, and since firms were, in effect, rationed as to what they could sell households were rationed as to how much labour they could supply in an economy in which markets were not clearing.

There is actually more common ground between these views than is sometimes supposed, for Benjamin and Kochin acknowledge the presence of demand deficiency in the depression years of the early 1930s, and Hatton sees unemployment benefits promoting layoffs and systematic short-time working as a rational response in a sales-constrained economy. Nevertheless, there is a crucial difference between the two views in terms of the potential role for economic policy between the wars. Hatton and Broadberry would claim that policies to increase government spending in particular, or aggregate demand in general, could have permanently reduced unemployment even in non-depression years, whereas Benjamin and Kochin would see such policies as unnecessary or even harmful in an economy which tended to automatic self-stabilization at an equilibrium (natural rate) of unemployment.

The economists' debate has, however, been largely conducted at the macroeconomic level, and the long-run structural story so popular among economic historians has not been well integrated into the argument. Not surprisingly, therefore, the economic historians complain that the economists' analysis is incomplete and ahistorical: 'a very substantial hard core of non-cyclical unemployment persisted throughout the period and the simple Keynesian approach to interwar unemployment has accorded insufficient attention to this essentially structural and regional problem' (Glynn and Booth, 1983, p. 335). The implication of this argument presumably would be that attempts to stimulate aggregate demand, even if necessary as a solution to the unemployment problem, would not have been sufficient because of the legacy of nineteenth-century industrialization, a view with which Keynes himself sympathized (Peden, 1980).

There are in principle, however, other ways of integrating the structural legacy with either a neo-classical or a Keynesian view of interwar unemployment. For example, a neo-classical argument might see the period as one of substantial structural change leading to a high turnover of jobs which in turn would create a high volume of voluntary 'search' unemployment among individuals transitionally out of work but sampling job offers. Alternatively, a Keynesian view might see the large staples sector experiencing an exogenous decline in its fortunes and introducing (particularly through a fall in exports) demand deficiency into the system. Evidently it is possible to have a more eclectic view embracing both these features, and it is certainly possible to infer such a position from the economic history literature (for example, Aldcroft, 1970, pp. 133–6), although it must be recognized that the connections between the structural legacy and interwar unemployment have not been made on the basis of explicit theory.

Indeed, the commonest approach seems to have been to infer that a 'structural problem' existed from the descriptive evidence on regional and industrial variations in unemployment. A selection of these data are given in Table 8.4. It should be remembered that the unemployment figures given here are for 'insured workers', and that the figures are not directly comparable with those for post World War II; it has been suggested that a multiplier of 8/13 would be required to roughly adjust the interwar figures to be comparable with later ones (Metcalf *et al.*, 1982). Moreover, regional differences are undoubtedly understated by the insured workers' evidence.

Table 8.4 highlights the uneven experience of unemployment between the wars. Obviously enough the South and Midlands have much lower levels of unemployment in each year, whether peaks (1929, 1937) or slump (1932), and these regions generally also have much lower shares of employment among their insured workers in the old staples, which are seen to have high recorded rates of unemployment among their insured workers. Nevertheless, Table 8.4 also reflects movement of labour, and it is particularly noticeable that fast-growing activities like building and road transport attracted so many extra workers that in 1937 recorded unemployment was higher than in 1929 and higher than in cotton textiles.

This suggestion of labour mobility prompts an obvious comment on the kind of evidence given in Table 8.4, and the danger of trying to infer too much from it. Of itself Table 8.4 certainly cannot demonstrate why unemployment *persisted* so long, nor answer the important question raised earlier as to the response of the economy to an increase in aggregate demand; nor, indeed, does the table fully pin down the connection between the early start and interwar unemployment.

V THE EARLY START AND STRUCTURAL UNEMPLOYMENT BETWEEN THE WARS

It is certainly the case that there is a tendency for the regions which were heavily specialized in the nineteenth-century staples to experience relatively high rates of both unemployment and structural change in employment during the interwar period. For counties there is a correlation of 0.77 between the unemployment rate in 1929 and the proportion of employment in textiles, mining, and shipbuilding taken together in 1921. Similarly, measuring structural change in employment by regions as $\sqrt{(\Sigma w_i(g_i - g)^2)}$, where w_i is the share in employment of the ith sector, g_i is the growth rate of employment in the ith sector, and g is the growth rate of regional employment,[1] shows for 1921–31 the North with a figure of 2.953 and Wales of 2.663 compared with the South East at 1.728 and East Anglia at 2.032. For comparison *no* region in 1901–11 had a figure greater than 1.608, or in 1951–61 greater than 2.009.

[1] This measure has been suggested in neo-classical inquiries into structural change, see Lilien (1982). All figures were calculated for the 27 sectors given in Lee (1979).

Perhaps more striking still is to note that reductions in employment in mining, iron and steel, shipbuilding, and textiles together between 1921 and 1931 were 8.2 per cent of the North West's total 1921 employment, 10.7 per cent of Wales's 1921 employment, and 13.8 per cent of the North's total employment, whereas negligible fractions of total employment were lost in these industries in the South East and East Anglia. The staple industries were indeed among the most localized of any employment categories (Lee, 1971, p. 233).

Over the period covered by the Censuses of Production, 1924–35, a rather similar story emerges. Total falls in employment in contracting industries were 962,000, of which shipbuilding, coalmining, and traditional textiles contributed 741,000. A crude calculation from the Census of Production based on average product per worker suggests that in these three sectors exports accounted for the employment of 368,000 fewer people in 1935 than in 1924. These numbers are quite large when set against a total unemployment of insured workers of 1,250,000 in 1929 and 1,482,000 in 1937, the two business-cycle peak years.

Thus the staple industries transmit a considerable reduction in aggregate demand to the economy and are also a major part of rapid changes in the structure of employment in the regions, which could indeed be expected to lead to a high volume of search unemployment.

A further feature of this experience comes from the work of Maizels (1963). Maizels noted that after 1913 Britain's exports to semi-industrial countries were substantially reduced by the growth of import-substituting production, and that in this regard British exports were much more severely affected than those of other leading countries. Between 1913 and 1937 Maizels estimates that exports of British manufactures to semi-industrial countries were reduced by import substitution by $1.47 bn. (1955), whereas the United States and Germany experienced reductions of $0.50 bn. and $0.49 bn. respectively. The British economy, as we have already seen, was, of course, much more heavily involved in exporting to semi-industrial countries and in exporting textiles than were the other two countries.

It seems likely, therefore, that the unemployment problem of the interwar years was exacerbated by the pattern of exports which comparative advantage based on unskilled labour and old tech-

Table 8.4. Regional[a] and Industrial Variations in Interwar Unemployment

(i) Regional Unemployment Rates for Insured Workers (%)	*June 1929*	*June 1932*	*June 1937*
South East	4.4	12.3	4.9
East Anglia	5.8	15.4	6.9
South West	7.7	16.1	6.5
West Midlands	9.0	22.0	6.2
East Midlands	8.4	19.7	7.5
North West	12.7	25.3	12.7
Yorkshire and Humberside	11.5	27.0	12.3
North	15.2	36.4	16.8
Wales	18.2	37.4	20.7
Scotland	11.0	26.8	14.0
Great Britain	10.4	22.1	10.8

(ii) Unemployment Rates for Insured Workers in Selected Industries (%)	*1929*	*1932*	*1937*
Coalmining	19.0	34.5	16.1
Cotton Textiles	12.9	30.6	10.9
Shipbuilding	25.3	62.0	24.4
Motors, Cycles, and Aircraft	7.1	22.4	5.9
Electrical Engineering	4.6	16.8	3.1
Chemicals	6.5	17.3	6.8
Building	14.3	30.2	14.6
Distributive Trades	6.2	12.6	8.9
Road Transport	12.0	22.2	12.3

(iii). Shares of Employment in Certain Sectors by Region in 1921 (%)

	Mining & Quarrying	*Textiles*	*Shipbuilding*	*Vehicles*	*Chemicals*	*Electrical Engineering*
South East	0.1	0.6	1.4	2.1	1.4	1.4
East Anglia	0.2	1.1	0.6	1.3	0.5	0.3
South West	3.5	1.7	2.2	2.1	0.8	0.2
West Midlands	6.6	2.3	0.04	7.0	0.7	1.9
East Midlands	11.9	10.8	0.2	2.7	0.6	0.4
North West	4.5	23.3	1.8	1.6	2.5	1.5
Yorkshire and Humberside	10.0	18.2	0.6	1.5	1.8	0.4
North	22.6	0.7	9.6	0.8	1.3	0.4
Wales	29.7	0.4	1.6	0.8	0.9	0.2
Scotland	8.2	7.0	5.7	0.8	0.9	0.4
Great Britain	7.4	6.9	2.2	2.0	1.3	0.9

Sources: Regional unemployment rates are calculated from the data in Beck (1951), Tables 18 and 19; industrial unemployment is taken from Tomlinson (1978, pp. 76–7); proportions of employment in different sectors are derived from Lee (1979).

[a] Regions are defined as the standard regions used by Lee (1979).

nology had bequeathed. Put another way, it seems possible that an economy which had exhibited a pattern of accumulation more geared towards human capital and research and development before 1914 would have been less exposed both to problems of structural adjustment and of demand deficiency in the 1920s. It is also important to recognize that there were adjustment problems arising from special circumstances in coal and shipbuilding which had expanded very fast during the wartime decade. Thus, whilst reductions in employment in the two industries amounted to 548,000 between 1921 and 1931, employment in 1931 was only 26,000 less than in 1911. These points weaken the argument that there was any automatic link between early industrialization and interwar unemployment.

There are, however, more important reasons to doubt the idea that the early start, or even 'overcommitment' before 1914, were *per se* responsible for a serious problem of structural unemployment between the wars. Although changes in the pattern of international economic activity required adjustment after 1918, an economy with the properties envisaged by the neo-classical economists' paradigm would surely have been able to cope well enough. The symptoms of the adjustment process would no doubt have involved a high labour turnover and, for a while at least, a high volume of search unemployment, but there would have been no persistent demand deficiency and no particular reason to expect periods of unemployment for given individuals to be of very long duration.

What we know of the interwar economy seems to be inconsistent with these expectations. The following points are particularly notable:

(1) Recent macroeconomic approaches to modelling the labour market have found that the demand for labour depended on the level of output, and that the labour market did not clear as predicted by neo-classical models (Hatton, 1983; Broadberry, 1983).

(2) Increased aggregate demand created by the rearmament programme appears to have raised employment significantly without leading to any substantial inflation or resulting in a 'crowding out' of other forms of expenditure. Thomas (1983, p. 566) estimates that by 1938 the increase in armaments expenditure over 1935 had raised employment by 1,037,000.

(3) Unemployment in depressed regions is not fully explained by their industrial composition, and did tend to fall when demand and employment rose in the more prosperous regions. In particular, recent research has confirmed that *all* industries in the regions of 'Outer Britain' tended to have significantly higher unemployment rates than their counterparts in 'Inner Britain' (Hatton, 1982).

(4) There was indeed an active labour market with considerable turnover. For example, in the quarter ended December 1938 unemployment rose by 45,000 to 1,695,000, but 1,031,000 became employed while 1,076,000 became unemployed (Singer, 1939/40, p. 43). Nevertheless, a most important facet of the 1930s labour market was the appearance of a substantial number of long-term unemployed, especially in the depressed regions. Such a phenomenon does not tie in at all well with a neo-classical account, which would predict a high volume of voluntary short-term unemployment of a search kind, its duration slightly prolonged by the availability of the dole. Thus, even at the business-cycle peak in 1937 there were 290,000 workers who had been unemployed for over a year, and, as Table 8.5 shows, unemployment was experienced on average in long spells in 'Outer Britain'.

Table 8.5. Average Weeks Duration of Male Unemployment: An Experience-weighted Measure for 1938

Great Britain	77.8	North East	68.5
London	30.9	North West	81.6
South East	34.1	North	142.7
South West	40.1	Scotland	112.2
Midlands	60.6	Wales	123.9

Source: Derived from the Ministry of Labour Gazette (1938) following the method developed by Main (1981); the regions are those of the Ministry of Labour at the time, not the standard regions used in Table 8.4.

The upshot of this discussion is the following. Britain's early industrialization and subsequent specialization in international trade did heighten problems of structural adjustment in the interwar economy, and to that extent the economic historians' insistence on a long-run perspective is justified. Nevertheless, the crucial point to remember is that the economy does not appear to have had the neo-classical attributes (spontaneous flexibility) to cope with this situation, and that when in the mid 1930s *de facto*

Keynesian remedies were tried unemployment was reduced by them in an economy experiencing demand deficiency. It might even be argued that the legacy of early industrialization was to increase the scope for Keynesian policy action rather than to render it irrelevant. It should certainly be noted, however, that a fully effective policy would have had to eradicate long-term unemployment, the causes of which are imperfectly understood and which require more research.

VI A REVITALIZATION OF THE ECONOMY BETWEEN THE WARS?

In section 8.3 above I argued that it is possible to regard the British economy before 1914 as 'overcommitted', and that the experience of the early start had encouraged this outcome. As noted earlier, Richardson (1967) suggested that this state of affairs ceased in the interwar period and that the structural adjustment problems of this time were part of a healthy revitalization of the economy. How does this claim look in the context of our knowledge of Britain's earlier growth and development?

In fact, in many ways economic performance in the interwar period seems to be a continuation of earlier trends rather than a refreshing change. Notably, investment in domestic fixed capital formation continued to be a low fraction of domestic product, total factor-productivity growth recovered only to mid-nineteenth-century levels, labour productivity in manufacturing remained below that of other leading economies, research and development spending rose but stayed much lower than in the United States, structural change was not very rapid in manufacturing, and all this was reflected in a continuation of earlier patterns of comparative advantage.

The details of this experience set in the context of earlier performance are as follows:

(1) Home and foreign investment together averaged 10.9 per cent of GNP in the years 1925–37, compared with 14.6 per cent in 1874–1913 (Matthews *et al.*, 1982, p. 140). Table 3.6 revealed that the late nineteenth-century figure itself was 4–5 percentage points low relative to the European norm. There was also a slight fall in the rate of growth of the capital stock, to 1.8 per cent per year in 1924–37 compared with 2.0 per cent in 1831–60, as

shown in Table 4.2, and the same rate in 1856–1913. Since the rate of growth of the labour force rose to 1.5 per cent in 1924–37, compared with 0.9 per cent in 1856–1913, the growth of capital per worker fell back to be roughly on a par with 1831–60. Much interwar net investment was in housing (Buxton, 1975), and it is therefore not surprising that the capital stock in manufacturing grew particularly slowly, at only one per cent per year in 1924–37.

(2) Total factor-productivity growth for the economy as a whole resumed Victorian levels. Matthews *et al.* (1982, p. 229) estimate a value of 0.7 per cent per year in 1924–37, which reference to Table 8.1 shows to be characteristic of 1856–99 and a little slower than the mid-nineteenth-century peak. Disaggregation does suggest new developments in that total factor productivity growth in manufacturing and agriculture in 1924–37 are 1.9 per cent and 2.1 per cent respectively, rates which Table 8.1 reveals to have been well above anything achieved in the nineteenth century. Nevertheless, the economy as a whole did not achieve the rapid rates of growth of overall productivity that Abramovitz and David (1973) found for the United States.

(3) Not surprisingly, therefore, labour productivity in industry remained below that of the United States. The estimates of Rostas (1948, p. 28) show output per worker in the United States 2.25 times that in Britain in 1937, compared with a ratio of about 1.7 in 1909, and Rostas also shows Germany to have output per worker 11 per cent higher than Britain in 1936. Thus, the tendency shown in Table 4.5 for labour productivity in British industry to be unimpressive by international standards for advanced economies continued after World War I.

(4) Although research and development spending grew considerably, thus marking a distinct change from the late nineteenth century, Britain continued to devote less resources to this activity than the United States. Sanderson estimated that spending on R&D had grown from less than £1 million per year in 1900–10 to somewhere between 0.15 and 0.29 per cent of GNP in the 1930s, but at the same time the United States's spending had risen to 0.5 per cent of GNP (1972b, p. 122). In 1937 revealed comparative advantage for the United States showed a strong correlation with the intensity of R&D, but Britain's did not (Crafts, 1985b).

(5) There was a shift towards newer industries; by 1937 chemicals, vehicles, and electrical engineering accounted for 21.1

per cent of manufacturing output, compared with 15.6 per cent in 1924 and only 8.8 per cent in 1900. Nevertheless, the overall rate of structural change in manufacturing in the interwar period was less than in 1913–24, 1937–51, or 1951–73 (Matthews *et al.*, 1982, pp. 255–7).

(6) It was noted above that the factor intensity of Britain's net exports remained remarkably similar right through to 1935, even though there were changes in the rank order of sectors' revealed comparative advantage. This suggests that Britain continued to accumulate human capital relatively slowly, and this also seems to be borne out by the work of Matthews *et al.* (1982, pp. 105–13, 260–6). Whilst the amount of formal schooling increased, so that the average years of schooling of a male worker stood at 8.14 years in 1931 compared with 6.75 years in 1911 and 4.69 years in 1881, technical training was seriously deficient (NEDO, 1983), and the movement of workers from low-paid to high-paid jobs slowed down such that the rate of quality shift in the labour force fell to only 0.05 per cent in 1924–37 compared with 0.1 per cent in 1871–1911 and 0.17 per cent in 1951–73.

Thus the pattern of growth between the wars seems still in many ways reminiscent of the Victorian epoch, and notions that the economy was revitalized between the wars should be tempered with a recognition that there were strong elements of similarity to the earlier age.

VII CONCLUSIONS

Britain's early and idiosyncratic industrialization did influence later growth and structural adjustment. What we have seen in this chapter is that when a new range of problems and opportunities opened up in the years after 1890 the economy seemed to lack the flexibility easily to respond, and that the earlier experience of industrialization probably made the required adjustments more difficult. Indeed, in terms of growth and international trade, if not unemployment, it is easy to point to substantial elements of continuity in performance, with the interwar economy still exhibiting familiar nineteenth-century patterns of low home investment and productivity growth and a non-technologically based comparative advantage.

All of this suggests that knowledge of Britain's early industrialization is important for an understanding of later economic history. Economic historians are right to warn of the dangers of ahistorical macroeconomic interpretations of the interwar economy. Nevertheless, we have seen in this chapter that the historical investigation and modelling necessary to establish the consequences of the early start has not yet progressed far enough to be embodied in a convincing economic analysis of growth and unemployment in the fifty years before World War II. This chapter has tried to encourage further thought and research in an important area; in the meantime the temptation to assert automatic consequences of the early start should be resisted.

Bibliography

Abramovitz, M. and David, P. A. (1973), 'Reinterpreting Economic Growth: Parables and Realities', *American Economic Review*, 63, Papers and Proceedings, 428–39.

Aldcroft, D. H. (1970), *The Interwar Economy: Britain, 1919–1939 (London: Batsford).*

Aldcroft, D. H. and Richardson, H. W. (1969), *The British Economy, 1870–1939* (London: Macmillan).

Allen, R. C. (1979), 'International Competition in Iron and Steel, 1850–1913', *Journal of Economic History*, 39, 911–37.

Allen, R. C. (1981), 'Entrepreneurship and Technical Progress in the Northeast Coast Pigiron Industry :1850–1913,' *Research in Economic History*, 6, 35–71.

Allen, R. C. (1982), 'The Efficiency and Distributional Consequences of Eighteenth Century Enclosures', *Economic Journal*, 92, 937–53.

André, D. (1971), *Indikatoren des Technischen Fortschritts. Eine Analyse der Wirtschaftsentwicklung in Deutschland von 1850–1913* (Göttingen).

Ashley, W. J. (1904), 'The Argument for Preference', *Economic Journal*, 14, 1–10.

Ashton, T. S. (1948), *The Industrial Revolution, 1760–1830* (London: Oxford University Press).

Ashton, T. S. (1949), 'The Standard of Life of the Workers in England, 1790–1830', *Journal of Economic History*, 9, 19–38.

Ashton, T. S. (1955), *An Economic History of England: The Eighteenth Century* (London: Methuen).

Bairoch, P. (1965), 'Niveaux de développement économique de 1810 à 1910', *Annales*, 20, 1091–117.

Bairoch, P. (1976), 'Europe's Gross National Product: 1800–1975', *Journal of European Economic History*, 5, 273–340.

Balassa, B. (1981), *The Newly Industrialising Countries in the World Economy* (Oxford: Pergamon Press).

Barrett-Brown, M. (1974), *The Economics of Imperialism* (Harmondsworth: Penguin).

Beck, G. M. (1951), *A Survey of British Employment and Unemployment, 1927–1945*, (Oxford: Institute of Economics).

Benjamin, D. and Kochin, L. (1979), 'Searching for an Explanation of Unemployment in Interwar Britain', *Journal of Political Economy*, 87, 441–70.

Benjamin, D. and Kochin, L. (1982), 'Unemployment and Unemployment Benefits in Twentieth Century Britain: A Reply to Our Critics', *Journal of Political Economy*, 90, 410–36.

Bennett, M. K. (1935), 'British Wheat Yield Per Acre for Seven Centuries', *Economic History*, 3, 12–29.

Berck, P. (1978), 'Hard Driving and Efficiency: Iron Production in 1890', *Journal of Economic History*, 38, 879–900.

Berend, I. and Ranki, G. (1980), 'Foreign Trade and the Industrialization of the European Periphery in the XIXth Century', *Journal of European Economic History*, 9, 539–84.

Beveridge, W. H. (1939), *Prices and Wages in England from the Twelfth to the Nineteenth Century*, vol. 1 (London: Longmans).

Bowley, A. L. (1900), *Wages in the United Kingdom in the Nineteenth Century* (Cambridge: Cambridge University Press).

Boyson, R. (1972), 'Industrialization and the Life of the Lancashire Factory Worker', in R. M. Hartwell (ed.), *The Long Debate on Poverty* (London: Institute of Economic Affairs).

Broadberry, S. N. (1983), 'Unemployment in Interwar Britain: A Disequilibrium Approach', *Oxford Economic Papers*, 35, 463–85.

Brown, J. (1983), 'Disamenities and the Standard of Living Debate: The Case of the Cotton Textile Workers of England, 1806–1850' (mimeo, University of Michigan).

Brownlee, J. (1926), 'History of the Birth and Death Rates in England and Wales taken as a Whole from 1570 to the Present Time', *Public Health*, 29, 211–22, 228–38.

Buxton, N. K. (1975), 'The Role of the "New" Industries in Britain during the 1930's: a Reinterpretation', (*Business History Review*), 49, 205–22.

Caron, F. (1979), *An Economic History of Modern France* (London: Methuen).

Caves, R. E. and Jones, R. W. (1981), *World Trade and Payments* (Boston, Mass.: Little, Brown).

Chambers, J. D. (1953), 'Enclosure and Labour Supply', *Economic History Review*, 5, 319–43.

Chambers, J. D. and Mingay, G. E. (1966), *The Agricultural Revolution, 1750–1880* (London: Batsford).

Chenery, H. B. and Syrquin, M. (1975), *Patterns of Development, 1950–1970* (London: Oxford University Press).

Cole, W. A. (1973), 'Eighteenth Century Economic Growth Revisited', *Explorations in Economic History*, 10, 327–48.

Cole, W. A. (1981), 'Factors in Demand, 1700–1780', in R. C. Floud and D. N. McCloskey (eds.), *The Economic History of Britain since 1700* (Cambridge: Cambridge University Press).

Coleman, D. C. (1956), 'Industrial Growth and Industrial Revolutions', *Economica*, 23, 1–22.

Coleman, D. C. (1958), *The British Paper Industry, 1495–1860* (Oxford: Clarendon Press).

Cottrell, P. L. (1980), *Industrial Finance, 1830–1914* (London: Methuen).

Crafts, N. F. R. (1976), 'English Economic Growth in the Eighteenth Century: A Re-examination of Deane and Cole's Estimates', *Economic History Review*, 29, 226–35.

Crafts, N. F. R. (1977), 'Industrial Revolution in Britain and France: Some Thoughts on the Question "Why Was England First?" ', *Economic History Review*, 30, 429–41.

Crafts, N. F. R. (1980a), 'Income Elasticities of Demand and the Release of Labour by Agriculture during the British Industrial Revolution', *Journal of European Economic History*, 9, 153–68.

Crafts, N. F. R. (1980b), 'National Income Estimates and the British Standard of Living Debate: A Reappraisal of 1801–1831', *Explorations in Economic History*, 17, 176–88.

Crafts, N. F. R. (1981), 'The Eighteenth Century: A Survey', in R. C. Floud and D. N. McCloskey (eds.), *The Economic History of Britain since 1700*, vol. 1 (Cambridge: Cambridge University Press).

Crafts, N. F. R. (1982), 'Regional Price Variations in England in 1843: An Aspect of the Standard of Living Debate', *Explorations in Economic History*, 19, 51–70.

Crafts, N. F. R. (1983a), 'British Economic Growth, 1700–1831: A Review of the Evidence', *Economic History Review*, 36, 177–99.

Crafts, N. F. R. (1983b), 'Gross National Product in Europe, 1879–1910: Some New Estimates', *Explorations in Economic History*, 20, 387–401.

Crafts, N. F. R. (1984a), 'Economic Growth in France and Britain, 1830–1910: A Review of the Evidence', *Journal of Economic History*, 44, 49–67.

Crafts, N. F. R. (1984b), 'A Time Series Study of Fertility in England and Wales, 1877–1938', *Journal of European Economic History*, 13, forthcoming.

Crafts, N. F. R. (1984c), 'Patterns of Development in Nineteenth Century Europe', *Oxford Economic Papers*, 36, 438–58..

Crafts, N. F. R. (1985a), 'English Workers' Real Wages during the Industrial Revolution: Some Remaining Problems', *Journal of Economic History*, 45, 139–44.

Crafts, N. F. R. (1985b), 'Revealed Comparative Advantage in Manufacturing, 1899–1950', *Journal of European Economic History*, 14, forthcoming.

Crafts, N. F. R. (1985c), 'Income Elasticities of Demand and the Release of Labor by Agriculture during the British Industrial Revolution: A Further Appraisal', in J. Mokyr (ed.), *The Economics of the Industrial Revolution*, (London: Allen and Unwin).

Crafts, N. F. R. and Thomas, M. (1984), 'Comparative Advantage

in UK Manufacturing, 1910–1935', (mimeo, University of Virginia).
Crouzet, F. (ed.) (1972), *Capital Formation in the Industrial Revolution* (London: Methuen).
Crouzet, F. (1980), 'Towards an Export Economy: British Exports during the Industrial Revolution', *Explorations in Economic History*, 17, 48–93.
Davies, D. (1795), *The Case of Labourers in Husbandry* (Bath).
Davis, R. (1969), 'English Overseas Trade, 1700–1774', in W. E. Minchinton (ed.), *The Growth of English Overseas Trade in the Seventeenth and Eighteenth Centuries* (London: Methuen), 99–120.
Davis, R. (1979), *The Industrial Revolution and British Overseas Trade* (Leicester: Leicester University Press).
Deane, P. (1957), 'The Output of the British Woollen Industry', *Journal of Economic History*, 17, 207–23.
Deane, P. (1961), 'Capital Formation in Britain before the Railway Age', *Economic Development and Cultural Change*, 9, 352–68.
Deane, P. (1968), 'New Estimates of Gross National Product for the United Kingdom, 1830–1914', *Review of Income and Wealth*, 14, 95–112.
Deane, P. (1973), 'The Role of Capital in the Industrial Revolution', *Explorations in Economic History*, 10, 349–64.
Deane, P. (1979), *The First Industrial Revolution* (Cambridge: Cambridge University Press).
Deane, P. and Cole, W. A. (1962), *British Economic Growth, 1688–1959* (Cambridge: Cambridge University Press).
Denison, E. F. (1967), *Why Growth Rates Differ* (Washington: Brookings).
Dixit, A. K. (1973), 'Theories of the Dual Economy: A Survey', in J. A. Mirrlees and N. H. Stern (eds.), *Models of Economic Growth* (New York: Wiley).
Edelstein, M. (1976), 'Realized Rates of Return on UK Home and Foreign Investment in the Age of High Imperialism', *Explorations in Economic History*, 13, 283–329.
Ellison, T. (1886), *The Cotton Trade of Great Britain* (London: Effingham Wilson).
Eversley, D. E. C. (1967), 'The Home Market and Economic Growth in England, 1750–1880', in E. L. Jones and G. E. Mingay (eds.), *Land, Labour and Population in the Industrial Revolution* (London: Arnold), 206–59.
FAO (1962), *Commodity Review (Special Supplement)* (Rome).
Farnie, D. A. (1979), *The English Cotton Industry and the World Market, 1815–1896* (Oxford: Oxford University Press).
Feinstein, C. H. (1972), *National Income, Expenditure and Output of the United Kingdom, 1855–1965* (Cambridge: Cambridge University Press).
Feinstein, C. H. (1978), 'Capital Formation in Great Britain', in P.

Mathias and M. M. Postan (eds.), *Cambridge Economic History of Europe*, vol. 7, part 1 (Cambridge: Cambridge University Press), 28–96.

Feinstein, C. H. (1981), 'Capital Accumulation and the Industrial Revolution', in R. C. Floud and D. N. McCloskey (eds.), *The Economic History of Britain since 1700*, vol. 1 (Cambridge: Cambridge University Press), 128–42.

Feinstein, C. H., *et al.* (1983), 'The Timing of the Climacteric and its Sectoral Incidence in the UK', in C. P. Kindleberger and G. di Tolla (eds.), *Economics in the Long View* (London: Macmillan), 168–85.

Floud, R. C. and McCloskey, D. N. (eds.) (1981), *The Economic History of Britain since 1700*, vol. 1 (Cambridge: Cambridge University Press).

Foreman-Peck, J. (1979), 'Tariff Protection and Economies of Scale: The British Motor Car Industry before 1939', *Oxford Economic Papers*, 31, 237–57.

Foreman-Peck, J. (1982), 'The American Challenge of the Twenties: Multinationals and the European Motor Industry', *Journal of Economic History*, 42, 865–81.

Fores, M. (1981), 'The Myth of a British Industrial Revolution', *History*, 66, 181–98.

Freudenberger, H. and Cummins, G. (1976), 'Health, Work and Leisure before the Industrial Revolution', *Explorations in Economic History*, 13, 1–12.

Fussell, G. E. (1929), 'Population and Wheat Production in the Eighteenth Century', *History Teachers' Miscellany*, 7, 65–111.

Gallmann, R. E. and Weiss, T. J. (1969), 'The Service Industries in the Nineteenth Century', in V. R. Fuchs (ed.), *Production and Productivity in the Service Industries* (New York: Columbia University Press).

Gayer, A. D. *et al.* (1953), *The Growth and Fluctuation of the British Economy, 1790–1850* (Oxford: Clarendon Press).

Glynn, S. and Booth, A. (1983), 'Unemployment in Interwar Britain: A Case for Relearning the Lessons of the 1930s', *Economic History Review*, 36, 329–48.

Gollop, F. M. and Jorgenson, D. W. (1980), 'US Productivity Growth by Industry, 1947–1973', in J. W. Kendrick and B. N. Vaccara (eds.), *New Developments in Productivity Measurement and Analysis* (Chicago: Chicago University Press).

Gourvish, T. (1972), 'The Cost of Living in Glasgow in the Early Nineteenth Century', *Economic History Review*, 25, 65–80.

Granger, C. W. J. and Elliott, C. M. (1967), 'A Fresh Look at Wheat Prices and Markets in the Eighteenth Century', *Economic History Review*, 20, 257–65.

Greenwood, M. J. and Thomas, L. B. (1973), 'Geographic Labor

Mobility in Nineteenth Century England and Wales', *Annals of Regional Science*, 7, 90–105.

Hammond, J. L. (1930), 'The Industrial Revolution and Discontent', *Economic History Review*, 2, 215–28.

Hannah, L. (1974), 'Takeover Bids in Britain before 1850: An Exercise in Business Pre-history', *Business History*, 16, 65–77.

Harley, C. K. (1982), 'British Industrialization before 1841: Evidence of Slower Growth during the Industrial Revolution', *Journal of Economic History*, 42, 267–89.

Harley, C. K. and McCloskey, D. N. (1981), 'Foreign Trade: Competition and the Expanding Economy', in R. C. Floud and D. N. McCloskey (eds.), *The Economic History of Britain since 1700*, vol. 2 (Cambridge: Cambridge University Press).

Harris, J. R. (1964), *The Copper King* (Liverpool: Liverpool University Press).

Hartwell, R. M. (1959), 'Interpretations of the Industrial Revolution in England: a Methodological Inquiry', *Journal of Economic History*, 19, 229–49.

Hartwell, R. M. (1961), 'The Rising Standard of Living in England, 1800–1850', *Economic History Review*, 13, 397–416.

Hartwell, R. M. (1967), 'Introduction', in R. M. Hartwell (ed.), *The Causes of the Industrial Revolution in England* (London: Methuen), 1–30.

Hartwell, R. M. (1971), *The Industrial Revolution and Economic Growth* (London: Methuen).

Hartwell, R. M. and Engerman, S. L. (1975), 'Models of Immiseration: the Theoretical Basis of Pessimism', in A. J. Taylor (ed.), *The Standard of Living in Britain in the Industrial Revolution* (London: Methuen), 189–213.

Hatton, T. J. (1982), 'Structural Aspects of Unemployment between the Wars' (mimeo, University of Essex).

Hatton, T. J. (1983), 'Unemployment Benefits and the Macroeconomics of the Interwar Labour Market: A Further Analysis', *Oxford Economic Papers*, 35, 486–505.

Hatton, T. J., *et al.* (1983), 'Eighteenth Century British Trade: Homespun or Empire Made', *Explorations in Economic History*, 20, 163–82.

Hawke, G. R. (1970), *Railways and Economic Growth in England and Wales, 1840–1870* (Oxford: Oxford University Press).

Hobsbawm, E. J. (1957), 'The British Standard of Living, 1790–1850', *Economic History Review*, 10, 46–68.

Hobsbawm, E. J. (1962), *The Age of Revolution, 1789–1848* (London: Weidenfeld and Nicolson).

Hobsbawm, E. J. (1963), 'The Standard of Living during the Industrial Revolution: A Discussion', *Economic History Review*, 16, 119–34.

Hobsbawm, E. J. (1968), Industry and Empire (London: Weidenfeld and Nicolson).

Hoffmann, W. G. (1955), *British Industry, 1700–1950* (Oxford: Blackwell).

Holderness, B. (1978), 'Productivity Trends in English Agriculture, 1600–1850: Observations and Preliminary Results', paper presented to the International Economic History Congress at Edinburgh.

Hueckel, G. (1981), 'Agriculture during Industrialisation', in R. C. Floud and D. N. McCloskey (eds.), *The Economic History of Britain since 1700*, vol. 1 (Cambridge: Cambridge University Press).

Hyde, C. K. (1977), *Technological Change and the British Iron Industry, 1700–1870* (Princeton, NJ: Princeton University Press).

Imlah, A. (1958), *Economic Elements in the Pax Britannica* (Cambridge, Mass.: Harvard University Press).

Inglis, B. (1971), *Poverty and the Industrial Revolution* (London: Hodder and Stoughton).

Ippolito, R. (1975), 'The Effect of the Agricultural Depression on Industrial Demand in England', *Economica*, 42, 298–312.

Jevons, W. S. (1965), *The Coal Question* (New York: Kelley).

John, A. H. (1965), 'Agricultural Productivity and Economic Growth in England, 1700–1760', *Journal of Economic History*, 25, 1–18.

Jones, E. L. (1974), *Agriculture and the Industrial Revolution* (Oxford: Blackwell).

Jones, E. L. (1981), 'Agriculture, 1700–1780', in R. C. Floud and D. N. McCloskey (eds.), *The Economic History of Britain since 1700*, vol. 1 (Cambridge: Cambridge University Press), 66–86.

Jones, E. L. and Woolf, S. J. (1969), 'The Historical Role of Agrarian Change in Economic Development', in E. L. Jones and S. J. Woolf (eds.), *Agrarian Change and Economic Development* (London: Methuen), 1–21.

Kanefsky, J. (1979), 'Motive Power in British Industry and the Accuracy of the 1870 Factory Return', *Economic History Review*, 32, 360–75.

Kelley, A. C. and Williamson, J. G. (1974), *Lessons from Japanese Development* (Chicago: Chicago University Press).

Kennedy, W. P. (1974), 'Foreign Investment, Trade and Growth in the United Kingdom, 1870–1913', *Explorations in Economic History*, 11, 415–44.

Kennedy, W. P. (1976), 'Institutional Reponse to Economic Growth: Capital Markets in Britain to 1914', in L. Hannah (ed.), *Management Strategy and Business Development* (London: MacMillan), 151–83.

Kennedy, W. P. (1982), 'Economic Growth and Structural Change in the United Kingdom, 1870–1914', *Journal of Economic History*, 42, 105–14.

Kravis, I. B., *et al.* (1978), 'Real GDP per Capita for More than One Hundred Countries', *Economic Journal*, 88, 215–42.

Kuznets, S. S. (1966), *Modern Economic Growth: Rate, Structure and Spread* (New Haven, Conn.: Yale University Press).

Kuznets, S. S. (1967), 'Quantitative Aspects of the Economic Growth of Nations: X: Level and Structure of Foreign Trade: Long Term Trends', *Economic Development and Cultural Change*, 15, 1–139.

Kuznets, S. S. (1971), *Economic Growth of Nations* (Cambridge, Mass.: Belknap).

Landes, D. S. (1969), *The Unbound Prometheus* (Cambridge: Cambridge Univesity Press).

Lawes, J. B. and Gilbert, J. H. (1880), 'On the Home Produce, Imports, Consumption and Price of Wheat', *Journal of the Statistical Society*, 43, 313–31.

Lee, C. H. (1971), *Regional Economic Growth in the United Kingdom since the 1880s* (London: Batsford).

Lee, C. H. (1979), *British Regional Employment Statistics, 1841–1971* (Cambridge: Cambridge University Press).

Lee, R. D. (1980), 'A Historical Perspective on Economic Aspects of the Population Explosion: The Case of Preindustrial England', in R. A. Easterlin (ed.), *Population and Economic Change in Developing Countries* (Chicago: University of Chicago Press), 517–57.

Lemon, C. (1838), 'The Statistics of the Copper Mines of Cornwall', *Journal of the Statistical Society*, 1, 65–84.

Lilien, D. 'Sectoral Shifts and Cyclical Unemployment', *Journal of Political Economy*, 90, 777–93.

Lindert, P. H. (1980), 'English Occupations, 1670–1811', *Journal of Economic History*, 40, 685–712.

Lindert, P. H. and Williamson, J. G. (1980), 'English Workers' Living Standards during the Industrial Revolution: A New Look', (mimeo, University of Wisconsin).

Lindert, P. H. and Williamson, J. G., (1982), 'Revising England's Social Tables 1688–1812', *Explorations in Economic History*, 19, 385–408.

Lindert, P. H. and Williamson, J. G. (1983a), 'Reinterpreting Britain's Social Tables, 1688–1913', *Explorations in Economic History*, 20, 94–109.

Lindert, P. H. and Williamson, J. G. (1983b), 'English Workers' Living Standards during the Industrial Rvolution: A New Look', *Economic History Review*, 36, 1–25.

McKeown, T. (1976), *The Modern Rise of Population* (London) Arnold).

McCloskey, D. N. (1970), 'Did Victorian Britain Fail?', *Economic History Review*, 23, 446–59.

McCloskey, D. N. (1981a), 'The Industrial Revolution: A Survey', in R. C. Floud and D. N. McCloskey (eds.), *The Economic History of Britain since 1700*, vol. 1 (Cambridge: Cambridge University Press), 103–27.

McCloskey, D. N. (1981b), *Enterprise and Trade in Victorian Britain* (London: Allen and Unwin).

Maddison, A. (1983), *Phases of Capitalist Development* (Oxford: Oxford University Press).

Main, B. G. M. (1981), 'The Length of Employment and Unemployment in Great Britain', *Scottish Journal of Political Economy*, 28, 146–64.

Maizels, A. (1963), *Industrial Growth and World Trade* (Cambridge: Cambridge University Press).

Marczewski, J. (1965), 'Le produit physique de l'economie française de 1789 à 1913 (comparaison avec la Grande Bretagne)', *Cahiers de l'Institut Statistique Economique Appliqué*, 4, 7–154.

Mathias, P. (1959), *The Brewing Industry in England, 1700–1830* (Cambridge: Cambridge University Press).

Mathias, P. (1983), *The First Industrial Nation* (London: Methuen).

Matthews, R. C. O., *et al.* (1982), *British Economic Growth, 1856–1973* (Stanford, Ca.: Stanford University Press).

Metcalf, D., *et al.* (1982), 'Still Searching for an Explanation of Unemployment in Interwar Britain', *Journal of Political Economy*, 90, 386–99.

Mitchell, B. R. (1975), *Abstract of European Historical Statistics* (London: Macmillan).

Mitchell, B. R. and Deane, P. (1962), *Abstract of British Historical Statistics* (Cambridge: Cambridge University Press).

Moggridge, D. E. (1972), *British Monetary Policy, 1924–1931* (Cambridge: Cambridge University Press).

Mulhall, M. G. (1884), *A Dictionary of Statistics* (London: Routledge).

Musson, A. E. (1978), *The Growth of British Industry* (London: Batsford).

Musson, A. E. (1982), 'The British Industrial Revolution', *History*, 67, 252–8.

NEDO (1983), *Education and Industry* (London: HMSO).

Nordhaus, W. D. and Tobin, J. (1972), *Is Growth Obsolete?* (New York: Columbia University Press).

O'Brien, P. K. (1982a), 'Long Swings in Relative Prices and the Significance of Agriculture for British Industrialization in the Eighteenth Century', (mimeo, University of Oxford).

O'Brien, P. K. (1982b), 'European Economic Development: The Contribution of the Periphery', *Economic History Review*, 35, 1–18.

O'Brien, P. K. and Engerman, S. L. (1981), 'Changes in Income and its Distribution during the Industrial Revolution', in R. C. Floud and D. N. McCloskey (eds.), *The Economic History of Britain since 1700*, vol. 1 (Cambridge: Cambridge University Press), 164–81.

O'Brien, P. K. and Keyder, C. (1978), *Economic Growth in Britain and France, 1780–1914* (London: Allen and Unwin).

Overton, M. (1979), 'Estimating Crop Yields from Probate Inventories: An Example from East Anglia, 1585–1735', *Journal of Economic History*, 39, 363–78.

Payne, P. (1974), *British Entrepreneurship in the Nineteenth Century* (London: Macmillan).

Peden, G. (1980), 'Keynes, The Treasury and Unemployment in the Later Nineteen-Thirties', *Oxford Economic Papers*, 32, 1–18.

Perkin, H. (1969), *The Origins of Modern English Society* (London: Routledge and Kegan Paul).

Phelps-Brown, E. H. and Hopkins, S. V. (1956), 'Seven Centuries of the Prices of Consumables, Compared with Builders' Wage Rates', *Economica*, 23, 296–314.

Pollard, S. (1980), 'A New Estimate of British Coal Production, 1750–1850', *Economic History Review*, 33, 212–35.

Pollard, S. (1981a), *Peaceful Conquest* (Oxford: Oxford University Press).

Pollard, S. (1981b), 'Sheffield and Sweet Auburn: Amenities and Living Standards in the British Industrial Revolution', *Journal of Economic History*, 41, 902–4.

Ricardo, D. (1817), *Principles of Political Economy and Taxation* (London: Murray).

Richardson, H. W. (1967), *Economic Recovery in Britain, 1932–1939* (London: Weidenfeld and Nicolson).

Riden, P. (1977), 'The Output of the British Iron Industry before 1870', *Economic History Review*, 30, 442–59.

Rostas, L. (1948), *Comparative Productivity in British and American Industry* (Cambridge: Cambridge University Press).

Rubinstein, W. D. (1981), *Men of Property* (London: Croom Helm).

Sandberg, L. G. (1974), *Lancashire in Decline* (Columbus, Ohio: Ohio State University Press).

Sandberg, L. G. (1981), 'The Entrepreneur and Technological Change', in R. C. Floud and D. N. McCloskey (eds.), *The Economic History of Britain since 1700*, vol. 2 (Cambridge: Cambridge University Press), 99–120.

Sanderson, M. (1972a), *The Universities and British Industry, 1850–1970* (London: Routledge and Kegan Paul).

Sanderson, M. (1972b), 'Research and the Firm in British Industry, 1919–1939', *Science Studies*, 3, 107–51.

Sanderson, M. (1983), *Education, Economic Change and Society in England, 1780–1870* (London: Macmillan).

Saul, S. B. (1965), 'The Export Economy, 1870–1914', *Yorkshire Bulletin of Economic and Social Research*, 17, 5–18.

Saul, S. B. (1968), *The Myth of the Great Depression, 1873–1896* (London: Macmillan).

Silberling, N. J. (1923), 'British Prices and Business Cycles, 1779–1850', *Review of Economic Statistics*, 5, 232–3.

Singer, H. W. (1939/40), 'Regional Labour Markets and the Process of Unemployment', *Review of Economic Studies*, 7, 42–58.

Smith, E. (1864), 'Appendix of the Sixth Report of the Medical Officer of the Privy Council', *British Parliamentary Papers*, 28, 216–329.

Spraos, J. (1983), *Inequalising Trade?* (Oxford: Clarendon Press).

Stern, R. M. and Maskus, K. E. (1981), 'Determinants of the Structure of US Foreign Trade, 1958–1976', *Journal of International Economics*, 11, 207–24.

Taylor, A. J. (ed.), *The Standard of Living in Britain in the Industrial Revolution* (London: Methuen).

Thomas, B. (1985), 'Food Supply in the United Kingdom During the Industrial Revolution', in J. Mokyr (ed.), *The Economics of the Industrial Revolution* (London: Allen and Unwin), 137–150.

Thomas, M. (1983), 'Rearmament and Economic Recovery in the Late 1930s', *Economic History Review*, 36, 552–79.

Thompson, E. P. (1963), *The Making of the English Working Class* (London: Gollancz).

Timmer, C. P. (1969), 'The Turnip, the New Husbandry and the English Agricultural Revolution', *Quarterly Journal of Economics*, 85, 375–95.

Tomlinson, J. (1978), 'Unemployment and Government Policy between the Wars: A Note', *Journal of Contemporary History*, 13, 65–78.

Tranter, N. (1981), 'The Labour Supply, 1780–1860', in R. C. Floud and D. N. McCloskey (eds.), *The Economic History of Britain since 1700*. vol. 1 (Cambridge: Cambridge University Press).

Tucker, R. S. (1936), 'Real Wages of Artisans in London, 1729–1935', *Journal of the American Statistical Association*, 31, 73–84.

Turner, M. E. (1982), 'Agricultural Productivity in England in the Eighteenth Century: Evidence from Crop Yields', *Economic History Review*, 35, 489–510.

Tyszynski, H. (1951), 'World Trade in Manufactured Commodities, 1899–1950', *Manchester School*, 19, 272–304.

Usher, D. (1980), *The Measurement of Economic Growth* (Oxford: Blackwell).

Von Tunzelmann, G. N. (1978), *Steam Power and British Industrialization to 1860* (Oxford: Oxford University Press).

Von Tunzelmann, G. N. (1981), 'Technical Progress during the Industrial Revolution', in R. C. Floud and D. N. McCloskey (eds.), *The Economic History of Britain since 1700* (Cambridge: Cambridge University Press), 143–63.

Von Tunzelmann, G. N. (1982), 'The Standard of Living, Investment and Economic Growth in England and Wales, 1760–1850', paper presented to the International Economic History Congress, Budapest.

Ward, J. R. (1974), *The Finance of Canal Building in Eighteenth Century England* (London: Oxford University Press).

Webb, S. (1980), 'Tariffs, Cartels, Technology and Growth in the German Steel Industry, 1879 to 1914', *Journal of Economic History*, 40, 309–29.

Williams, E. (1944), *Capitalism and Slavery* (Chapel Hill, NC: University of North Carolina Press).

Williams, J. E. (1966), 'The British Standard of Living, 1750–1850', *Economic History Review*, 19, 581–9.

Williamson, J. G. (1981), 'What do we Know about Skill Accumulation in Nineteenth Century Britain?' (mimeo, University of Wisconsin).

Williamson, J. G. (1982), 'Was the Industrial Revolution Worth It? Disamenities and Death in Nineteenth Century British Towns', *Explorations in Economic History*, 19, 221–45.

Williamson, J. G. (1984), 'British Mortality and the Value of Life, 1781–1931', *Population Studies*, 38, 157–72.

Woodward, D. (1971), 'Agricultural Revolution in England, 1500–1900: A Survey', *The Local Historian*, 9, 323–33.

Wrigley, E. A. and Schofield, R. (1981), *The Population History of England, 1541–1871* (London: Arnold).

Index